ENERGY SCIENCE, ENGINEERING AND TECHNOLOGY

IMPROVING THE EFFICIENCY OF COAL-FIRED POWER PLANTS

ISSUES AND POTENTIAL BENEFITS

ENERGY SCIENCE, ENGINEERING AND TECHNOLOGY

Additional books in this series can be found on Nova's website under the Series tab.

Additional e-books in this series can be found on Nova's website under the e-book tab.

ENERGY SCIENCE, ENGINEERING AND TECHNOLOGY

IMPROVING THE EFFICIENCY OF COAL-FIRED POWER PLANTS

ISSUES AND POTENTIAL BENEFITS

ROLAND LUDOVIC
EDITOR

NOVINKA

New York

For permission to use material from this book please contact us:
Telephone 631-231-7269; Fax 631-231-8175
Web Site: http://www.novapublishers.com

LIBRARY OF CONGRESS CATALOGING-IN-PUBLICATION DATA

ISBN: 978-1-63117-590-9

Published by Nova Science Publishers, Inc. † New York

CONTENTS

PREFACE

This book focuses on efficiency improvements to power plants, and discusses retrofits, technologies, and other modifications to facility operations which offer the potential to improve power plant efficiency and reduce CO2 emissions.

Chapter 1 – Coal has long been the major fossil fuel used to produce electricity. However, coal-fired electric power plants are one of the largest sources of air pollution in the United States, with greenhouse gas (GHG) emissions from burning of fossil fuels believed to be the major contributor to global climate change. Regulations under development at the Environmental Protection Agency (EPA) would impose new requirements on fossil-fueled (mostly coal-fired) power plants (CFPPs) to control GHG emissions. The first of these requirements was issued in September 2013 with proposed standards for the control of carbon dioxide (CO_2) emissions from new electric generating units burning fossil fuels. EPA's proposals for control of GHG emissions from existing power plants are expected by June 2014, with many options under consideration. EPA may target emissions on a state or plant-by-plant basis, with companies likely given choices for compliance. Within such a system, efficiency improvements can be an important contributor.

The overall efficiency of a power plant encompasses the efficiency of the various components of a generating unit. Minimizing heat losses is the greatest factor affecting the loss of CFPP efficiency, and there are many areas of potential heat losses in a power plant. Efficiency of older CFPPs becomes degraded over time, and lower power plant efficiency results in more CO_2 being emitted per unit of electricity generated.

The options most often considered for increasing the efficiency of CFPPs include equipment refurbishment, plant upgrades, and improved operations and maintenance schedules.

Cost of the improvements is often compared to the expected return in increased efficiency as a primary determinant of whether to go forward with a program. A study by the Asia-Pacific Working Group (APWG) found that at the low to medium end of cost expenditures are combustion, steam cycle, and operations and maintenance improvements. Replacing the older CFPPs with new power plants was not generally seen as being practical because the expenditure for a new plant could not be justified by the improved performance. Instead, efficiency and operational improvements were seen as a possible alternative considering a range of equipment upgrades and refurbishment options to various CFPP systems.

The National Energy Technology Laboratory (NETL) took APWG's analysis a step further, finding that while the average efficiency of U.S. plants was 32% in 2007, the efficiency of the top 10% was five points higher at 37.4%. NETL suggested that if GHG emissions reduction was a goal, then heat rate efficiency improvements could enable a power plant to generate the same amount of electricity from less fuel and decrease CO_2 emissions.

In 2010, NETL completed a new study of U.S. CFPP efficiency, concluding that if generation levels were held constant at 2008 levels, overall fleet efficiency could be raised from 32% to 36%, resulting in an overall reduction in U.S. GHG emissions of 175 million metric tonnes per year, or 2.5% of total U.S. GHG emissions in 2008.

According to subsequent analyses, NETL concluded that retirements of lower efficiency units combined with increased generation from higher efficiency refurbished units, and advanced refurbishments with improved operation and maintenance, would be necessary to achieve this goal. These improvements would generally be considered low to medium cost upgrades.

However, at the higher cost end are major plant retrofits and upgrades (i.e., conversion of subcritical CFPP units to super- or ultra-supercritical CFPP units), which would raise efficiencies more substantially.

One possible approach to achieve fleet-wide efficiency improvement might be to follow NETL's suggestion of using the top decile of CFPP efficiency as a benchmark for the U.S. fleet, and establish an "efficiency frontier" that would be revisited periodically to reset the benchmark. This could be combined with possible incentives to improve efficiency or retire less efficient power plants.

Other federal approaches could use tax incentives to encourage greater efficiency, or employ energy efficiency standards focused on improving efficiency of CFPPs. The overall cost of these or other programs to increase CFPP efficiency has yet to be determined.

Chapter 2 – In 2008, the U.S. coal-fired power plant (CFPP) fleet had a generation-weighted average efficiency of 32.5% while the top ten percent of the fleet had an efficiency of 37.6%, five percentage points higher. The generating units in the top ten percent are diverse (they are not all new, large, super critical plants), indicating an opportunity for fleet-wide efficiency improvement. The National Energy Technology Laboratory (NETL) segmented the fleet into 13 groups based on characteristics that limit efficiency, and calculated the best-in-class efficiency within each group. Based on each group achieving an average efficiency equal to its 90th percentile, the overall CFPP fleet average efficiency would be 35.2%. NETL sets forth a vision of 36% based on retirements of low efficiency units, and improvements within the best-in-class. Under a scenario where generation from coal is constant at the 2008 level, increasing the average efficiency from 32.5% to 36% reduces U.S. GHG by 175 MMmt/year [1] or 2.5% of total U.S. GHG emissions in 2008.

Chapter 3 – This document is one of several white papers that summarize readily available information on control techniques and measures to mitigate greenhouse gas (GHG) emissions from specific industrial sectors. These white papers are solely intended to provide basic information on GHG control technologies and reduction measures in order to assist States and local air pollution control agencies, tribal authorities, and regulated entities in implementing technologies or measures to reduce GHGs under the Clean Air Act, particularly in permitting under the prevention of significant deterioration (PSD) program and the assessment of best available control technology (BACT). These white papers do not set policy, standards or otherwise establish any binding requirements; such requirements are contained in the applicable EPA regulations and approved state implementation plans.

This document provides information on control techniques and measures that are available to mitigate GHG emissions from the coal-fired electric generating sector at this time. The primary GHG emitted by the coal-fired electric generation industry is carbon dioxide (CO_2), and the control technologies and measures presented in this document focus on this pollutant. While a large number of available technologies are discussed here, this paper does not necessarily represent all potentially available technologies or measures that that may be considered for any given source for the purposes of reducing its GHG emissions. For example, controls that are applied to other industrial source categories with exhaust streams similar to the cement manufacturing sector may be available through "technology transfer" or new technologies may be developed for use in this sector.

In: Improving the Efficiency ...
Editor: Roland Ludovic

ISBN: 978-1-63117-590-9

Chapter 1

INCREASING THE EFFICIENCY OF EXISTING COAL-FIRED POWER PLANTS*

Richard J. Campbell

SUMMARY

Coal has long been the major fossil fuel used to produce electricity. However, coal-fired electric power plants are one of the largest sources of air pollution in the United States, with greenhouse gas (GHG) emissions from burning of fossil fuels believed to be the major contributor to global climate change. Regulations under development at the Environmental Protection Agency (EPA) would impose new requirements on fossil-fueled (mostly coal-fired) power plants (CFPPs) to control GHG emissions. The first of these requirements was issued in September 2013 with proposed standards for the control of carbon dioxide (CO_2) emissions from new electric generating units burning fossil fuels. EPA's proposals for control of GHG emissions from existing power plants are expected by June 2014, with many options under consideration. EPA may target emissions on a state or plant-by-plant basis, with companies likely given choices for compliance. Within such a system, efficiency improvements can be an important contributor.

The overall efficiency of a power plant encompasses the efficiency of the various components of a generating unit. Minimizing heat losses is the greatest factor affecting the loss of CFPP efficiency, and there are

* This is an edited, reformatted and augmented version of Congressional Research Service Publication, No. R43343, dated December 20, 2013.

many areas of potential heat losses in a power plant. Efficiency of older CFPPs becomes degraded over time, and lower power plant efficiency results in more CO_2 being emitted per unit of electricity generated. The options most often considered for increasing the efficiency of CFPPs include equipment refurbishment, plant upgrades, and improved operations and maintenance schedules.

Cost of the improvements is often compared to the expected return in increased efficiency as a primary determinant of whether to go forward with a program. A study by the Asia-Pacific Working Group (APWG) found that at the low to medium end of cost expenditures are combustion, steam cycle, and operations and maintenance improvements. Replacing the older CFPPs with new power plants was not generally seen as being practical because the expenditure for a new plant could not be justified by the improved performance. Instead, efficiency and operational improvements were seen as a possible alternative considering a range of equipment upgrades and refurbishment options to various CFPP systems.

The National Energy Technology Laboratory (NETL) took APWG's analysis a step further, finding that while the average efficiency of U.S. plants was 32% in 2007, the efficiency of the top 10% was five points higher at 37.4%. NETL suggested that if GHG emissions reduction was a goal, then heat rate efficiency improvements could enable a power plant to generate the same amount of electricity from less fuel and decrease CO_2 emissions.

In 2010, NETL completed a new study of U.S. CFPP efficiency, concluding that if generation levels were held constant at 2008 levels, overall fleet efficiency could be raised from 32% to 36%, resulting in an overall reduction in U.S. GHG emissions of 175 million metric tonnes per year, or 2.5% of total U.S. GHG emissions in 2008.

According to subsequent analyses, NETL concluded that retirements of lower efficiency units combined with increased generation from higher efficiency refurbished units, and advanced refurbishments with improved operation and maintenance, would be necessary to achieve this goal. These improvements would generally be considered low to medium cost upgrades.

However, at the higher cost end are major plant retrofits and upgrades (i.e., conversion of subcritical CFPP units to super- or ultra-supercritical CFPP units), which would raise efficiencies more substantially.

One possible approach to achieve fleet-wide efficiency improvement might be to follow NETL's suggestion of using the top decile of CFPP efficiency as a benchmark for the U.S. fleet, and establish an "efficiency frontier" that would be revisited periodically to reset the benchmark. This could be combined with possible incentives to improve efficiency or retire less efficient power plants.

Other federal approaches could use tax incentives to encourage greater efficiency, or employ energy efficiency standards focused on improving efficiency of CFPPs. The overall cost of these or other programs to increase CFPP efficiency has yet to be determined.

INTRODUCTION

Coal has long been the major fossil fuel used to produce electricity. However, the Environmental Protection Agency (EPA) lists coal-fired electric power plants as one of the largest sources of air pollution in the United States, with greenhouse gas[1] (GHG) emissions from burning fossil fuels believed to be the largest contributor to global climate change.

Regulations under development at EPA would impose new requirements on power plants to control GHG emissions. First, in September 2013 EPA proposed standards for the control of carbon dioxide (CO_2) emissions from new electric generating units burning fossil fuels. EPA has suggested that utilization of carbon capture and storage (CCS) is a viable means for new coal-fired power plants to comply with the proposed standards.[2] But higher efficiency components and processes are unlikely to be sufficient to meet the proposed new plant standards.

As requirements for new sources (i.e., new power plants), EPA's proposed standards do not directly apply to existing power plants currently producing electricity. EPA's proposals for control of GHG emissions from existing power plants are expected by June 2014,[3] with many options for reducing GHGs under consideration. EPA may target emissions on a state or plant-by-plant basis, with companies likely given choices for compliance, and increasing coal-fired power plant (CFPP) efficiency may be one of those choices.

Improving the efficiency of existing coal plants could potentially result in significant reductions of CO_2 emissions per unit of electricity produced. However, certain modifications to power plants to increase power output can potentially increase pollutant emissions, thus triggering new source review[4] (NSR) requirements.

Therefore, any modifications made must be shown to reduce pollutants if NSR is to be avoided. Expenditures to increase efficiency would likely be evaluated on a cost vs. benefits approach, with modifications to improve efficiency varying according to many factors, including the type of fuel burned, and the age and the physical condition of the power plant.

Carbon capture and sequestration (CCS) will not be a focus of improvements discussed in this report, as there are no CCS technologies considered as commercially available for full-scale application to the broad majority of existing coal-fired power plants,[5] and EPA has stated that it does not expect to require CCS at *existing* plants.

This report focuses on efficiency improvements to power plants, and discusses retrofits, technologies, and other modifications to facility operations which offer the potential to improve power plant efficiency and reduce CO_2 emissions.

Some in Congress have expressed concerns about the potential impacts on electricity reliability and fuel diversity from retirements of coal plants due to pending and new environmental regulations. Increasing efficiency of coal plants may help to address these concerns by reducing emissions without reducing output. Additionally, Congress may want to consider whether such efficiency improvements could be accelerated if these were implemented in a program focused on increasing the efficiency of the coal-fired power plant sector.

COAL AND EXISTING U.S. COAL POWER PLANTS

For most of the history of power generation in the United States, coal has been the dominant fuel used to produce electricity. In 2012, coal was used to fuel approximately 38% of power generation in the United States, as shown in **Figure 1**. Coal has been the fuel of choice for many decades because of its wide availability, and the relatively low cost of producing electricity in large, coal-burning power plants. Coal's low-priced, high energy content enabled the building of power plants able to take advantage of economies of scale in steam-electric production.

In a steam power plant, coal (or other combustible fuel) is burned to provide heat for turning water into steam in a boiler.[6] The steam is then forced under pressure into a steam turbine-driven generator which produces electricity. As of 2012, the U.S. coal-powered generation fleet consisted of 1,337 units[7] with a nameplate capacity of almost 313 gigaWatts (GW) of generating capacity.[8]

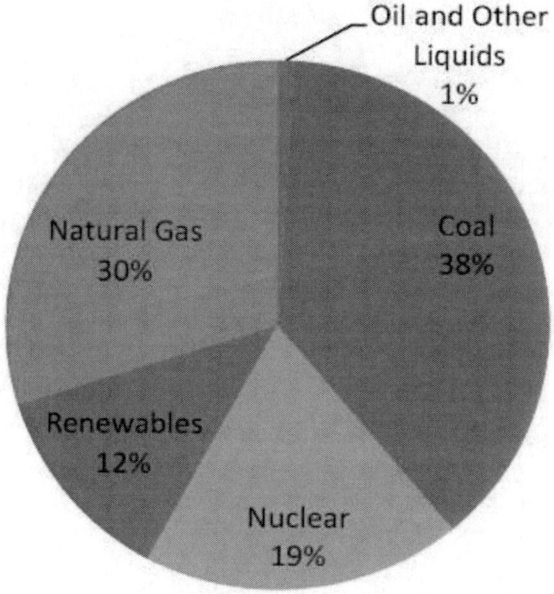

Source: U.S. Energy Information Administration, Electric Power Monthly, September 2013.

Figure 1. Electricity Generation by Fuel, 2012.

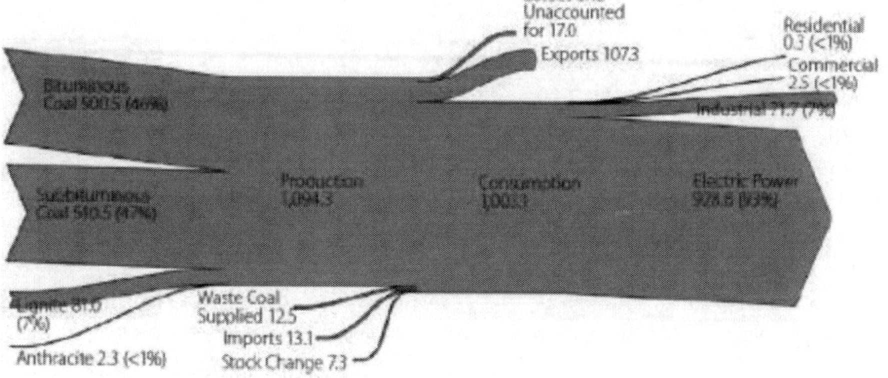

Source: U.S. Energy Information Administration. See http://www.eia.gov/totalenergy/data/annual/pdf/sec7_3.pdf.

Figure 2. Flow of U.S. Coal Consumption for 2011; (Million Short Tons).

Coal and Greenhouse Gas Emissions

Coal is largely composed of carbon, hydrogen and oxygen, with varying amounts of carbon, sulfur, ash, and moisture content in the different types of coal mined in the United States. **Figure 2** shows the use of the four major types (also called "ranks") of coal produced in the United States, with bituminous and subbituminous coal dominating electric power generation.

> **Bituminous** is the most abundant form of coal in the United States, and is the type most commonly used to generate electricity. Bituminous coal has a carbon content ranging from 45% to 86%, and a heat value between 10,500 British Thermal Units (BTUs)[9] and 15,500 BTUs per pound.
>
> **Subbituminous** coal is mostly found in six western states and Alaska. It has a carbon content of between 35% and 45%, and a heat value of between 8,300 BTUs and 13,000 BTUs. Subbituminous coal generally has a lower sulfur content than other types of coal.
>
> **Lignite** has the lowest carbon content of the four types of coal generally used for electric power generation, averaging between 25% and 35%, and a high moisture and ash content. It also has the lowest heat value, ranging between 4,000 BTUs and 8,300 BTUs.
>
> **Anthracite** has the highest carbon content (between 86% and 98%), and a heat value of about 15,000 BTUs. Anthracite coal is a small part of the electric power market, and is mostly found in the Appalachian region of Pennsylvania.

The amount of CO_2 emitted from a coal-fired power plant (CFPP) varies with the type of coal burned. The combustion of coal in the presence of oxygen causes its carbon and hydrogen constituents to react, releasing CO_2 emissions and water, with varying amounts of other products such as oxides of nitrogen and sulfur, carbon monoxide, and fine particulate matter. Carbon dioxide is the primary emission of concern when considering GHG emissions from power plants.[10]

Generally, anthracite emits the largest amount of CO_2 per million BTUs (MMBTU) of coal burned, followed by lignite, subbituminous coal, and bituminous coal. Carbon dioxide emissions from coal-fired power plants could thus be reduced by burning a better grade of coal, or by increasing the efficiency of the power plant and reducing overall coal consumption, without a need to completely repower the plant (in, say, a coal to natural gas conversion). Fuel switching may be necessary if a greater degree of CO_2 emissions reduction is desired.

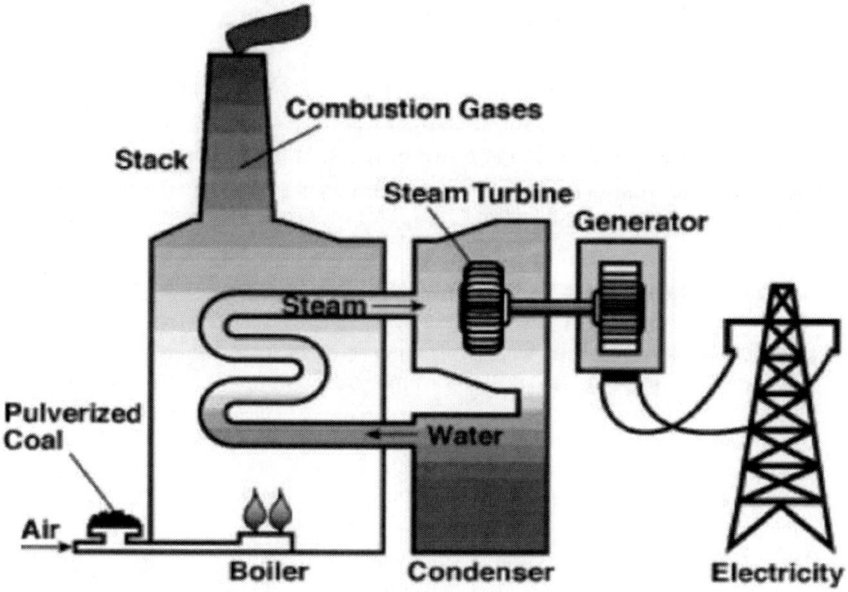

Source: Oncor Electric Delivery Company. Steam Turbine.
Notes: See http://www.c2es.org/technology/overview/electricity.

Figure 3. Electric Power Generation; Steam Turbine- Generator.

Types of U.S. Coal-Fired Power Plants

Steam turbines are at the heart of coal-fired power plants. As shown in the simplified schematic of a pulverized coal plant in **Figure 3**, a steam electric power plant consists of a number of basic components. Coal is crushed and fed into a boiler where it is burned to heat water into steam. The steam is injected under pressure into a turbine which turns a generator (where essentially a magnet turns in a coil of wire causing electrons to flow thus creating an electric current). Steam returning from the turbine is then cooled in a condenser, and the water is fed back by a feedwater pump to the boiler to continue the process. The expansion of water into steam vapor (and condensation back into liquid water) in this manner is called a Rankine Cycle,[11] and is the basis for most electric power generation in the United States.

A typical coal-fired power plant has multiple generating units, each with its own steam generating boiler. Usually, coal is pulverized by a combination of crushing and grinding until a desired degree of fineness is achieved. The coal is sieved, and dried using heated air before it is conveyed to a furnace where it is burned to produce steam. Steam pressure and temperature are specifically related, as steam's temperature rises with increasing steam pressure. The pressure and temperature of the steam produced have been rising steadily over the years, ranging from yesterday's sub-critical units to today's ultra-super critical units.

> **Subcritical steam** generation units operate at pressures such that water boils first and then is converted to superheated steam. At **supercritical** pressures, water is heated to produce superheated steam without boiling. Due to the improved thermodynamics of expanding higher pressure and temperature steam through the turbine, a **supercritical steam** generating unit is more efficient than a subcritical unit.
>
> **Ultra-supercritical steam** (USC) generation currently is the most efficient technology for producing electricity fueled by pulverized coal. A USC unit operates at supercritical pressure and at advanced steam temperatures of 1,100 °F ...These temperatures and pressures enable more efficient operation of the turbine cycle. This increase in efficiency reduces fuel (coal) consumption, and thereby reduces emissions, solid waste, water use and operating costs.[12]

While pulverized coal units are most common in the United States, coal-fired power plants use other technologies to burn coal including cyclone-fired boilers,[13] fluidized bed combustion,[14] and integrated coal gasification /combined cycle[15] technologies.

U.S. Coal Power Plants Are Aging

According to EIA, approximately 73% of U.S. coal-fired power plants were age 30 years or older at the end of 2010. The service life for CFPPs normally averages between 35 and 50 years, and varies according to boiler type, maintenance practices, and the type of coal burned, among other factors.

Vintages of existing power plants can be seen in **Figure 4**, which illustrates the age and capacity of coal-fired and other power plants. Most of this generating capacity was built between 1950 and 1990. The aging majority of U.S. coal-fired capacity contrasts against the recent surge in younger wind power and natural gas-fired capacity additions whose generating capacity is mostly less than 10 years old.

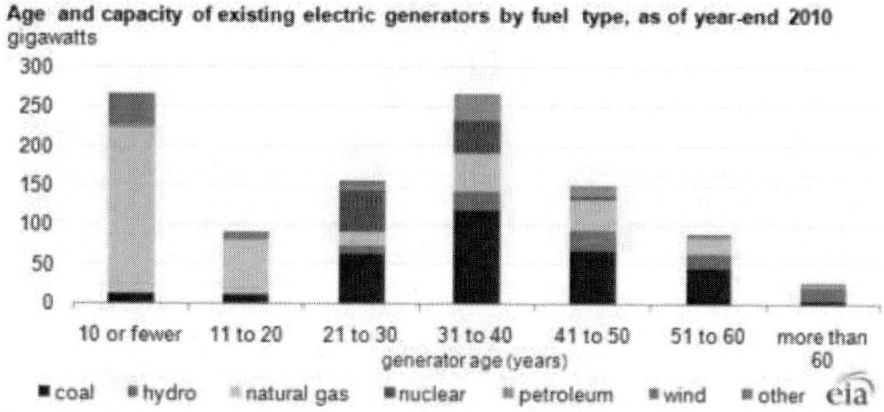

Source: U.S. Energy Information Administration. Today in Energy.
Notes: See http://www.eia.gov/todayinenergy/detail.cfm?id=1830.

Figure 4. Vintages of U.S. Generating Capacity; As of 2010.

The efficiency of coal-fired power plants, in particular, decreases with age. While good maintenance practices can keep power plant efficiency high in the early years of life, as the plant ages, power plant performance and efficiency erode after about 25 to 30 years of operation,[16] and substantial work may be required to keep the plant operating efficiently and economically.

All power plants are subject to retirement when they reach the end of their useful service life. As power plants age, they are generally upgraded to continue operations, but the least efficient plants may be retired. Other plants may be shifted from base load operations (in which they essentially operate around the clock) to less demanding intermediate or peaking schedules. The cost of building a power plant is generally recovered over the depreciable life of the asset, such that operations and maintenance (O&M) expenses become the major component of an older power plant's continuing costs. A major component of O&M is the cost of fuel, and the expectation of continued lower prices for natural gas is weighing on decisions concerning whether many older, less efficient coal power plants will be mothballed or closed altogether. The costs of modernizing older power plants to meet new regulatory requirements can be relatively high. When the cost of upgrades to meet new environmental requirements is considered along with (perhaps increasing) O&M expenses, many older coal power plants are likely to face outright retirement decisions.[17]

Efficiency of Power Plants and Power Plant Systems

Improving the efficiency of existing coal plants could potentially result in large reductions of CO_2 emissions per unit of electricity produced. Since GHG emissions from electricity generation are essentially composed of CO_2 emissions, improvements in efficiency are a direct means of reducing GHG emissions.

The performance of a power plant can be expressed by a number of measures, including heat rate (i.e., the efficiency of conversion from fuel energy input to electrical energy output), and thermal efficiency. In its simplest form, a plant's heat rate (for a particular period) can be defined as follows:

$$HR = F / E$$

where,

HR = heat rate (Btu/kWh)
F = heat energy input supplied by fuel to the power plant for a period (BTU)
E = energy output from the power plant in a period (kWh)

Since the equivalent BTU content of a single kWh of electricity is 3,412 BTU, thermal efficiency[18] can be calculated as:

$$TE = (100)\,(3412) / HR$$ *where,*
TE = thermal efficiency (%)

As an example, using the average heat rate in 2011 of 10,444 BTU/kWh for coal-fired power plants (i.e., all coal types), the average efficiency for coal-fired plants was 33%.[19]

A lower heat rate represents a more efficient generating unit, since it requires less heat input to generate a kWh of electric energy. A generating unit can thus improve its efficiency by reducing the fuel it uses relative to a specific amount of electricity generated, thus reducing the amount of CO_2 emitted.

> A percentage improvement in heat rate is nearly equivalent to an equal percentage improvement in the emissions rate in terms of the change in CO_2 emissions. The difference stems from the small variation in carbon per Btu across coal varieties. The heterogeneity in heat rates across coal-fired generation units can partly be explained by technical characteristics determined at the time of plant construction that cannot be changed without a major overhaul. This category includes size, age, firing type, and the technology employed. Higher efficiency is generally associated with plants that are used more heavily because efficient units are less costly to operate.
>
> A second factor is how the boiler is used. The relationship between the heat rate and utilization is nonlinear, as efficiency tends to be lower at very low and very high levels of utilization ... Units with lower utilization may be ramped up and down more frequently, which requires additional fuel input as temperature in the boiler fluctuates. The result could involve efficiency losses at least partly outside the control of plant decision makers. Plant managers control several other factors that affect heat rates. Techniques, management, or technology may improve the efficiency of the plant by targeting the major components of the coal combustion process: oxygen, temperature, and pressure. Excessive deviations in any of these areas may decrease efficiency through waste or shortfalls ...Maintenance and performance testing are also critical for identifying and preventing losses.[20]

Therefore, in practical terms, a power plant's heat rate can be affected by a number of factors and power plants systems. Heat rate may present one measure of efficiency, but when considering power plant GHG emissions, measuring carbon dioxide emissions per unit of energy output (i.e., per kWh or per MWh of generation) may provide a more useful measure.[21]

Each power plant thus presents a unique opportunity when looking at the issue of increasing efficiency, and reducing emissions. **Figure 5** illustrates a relationship between efficiency improvement and CO_2 emissions (for CFPPs using bituminous coal) highlighting the lower emissions of higher pressure CFPPs. Upgrading from subcritical operation to supercritical steam conditions (with required pollution technology) could add at least 20 years to a plant's service life,[22] depending on the regulatory and environmental regime in place. A subcritical plant could achieve at best 40% efficiency (on an LHV basis), while a supercritical steam plant could potentially achieve an efficiency two points higher and emit 4% less CO_2.[23] Advancing the technology from a supercritical to an advanced ultra-supercritical CFPP could see an efficiency of 46% to 48%, which could mean as much as 18% to 22% less CO_2 per MWh generated than an equivalent-sized subcritical PC unit.[24] However, "[m]ajor plant upgrading involving conversion of subcritical to supercritical or ultra-supercritical ... has seldom progressed beyond studies because of the high cost."[25]

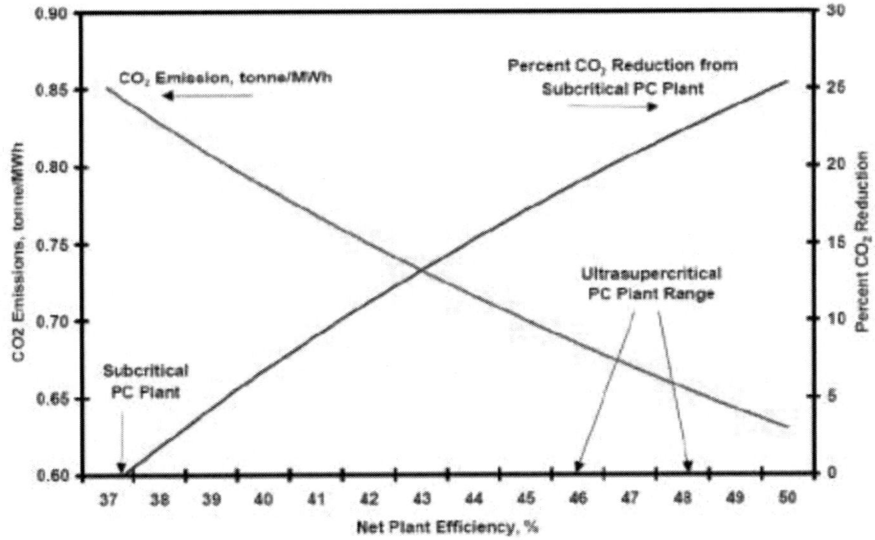

Source: Booras, G. and N. Holt, Pulverized Coal and IGCC plant Cost and Performance Estimates, Gasification Technologies Conference Washington D.C. 2004.

Notes:

1) One tonne (also known as a metric ton) is a unit of mass equaling 1,000 kilograms.
2) This chart assumes subcritical plants (using bituminous coal) would be near the top of the range for efficiency. Improvements at the higher end of the range are now considered to be based on advanced ultrasupercritical technology, which while technologically possible, is unlikely to be implemented in the next few years.

Figure 5. Carbon Dioxide Emissions vs. Net Plant Efficiency; Based on Burning Pittsburgh #8 (Bituminous) Coal.

EFFICIENCY IMPROVEMENTS TO REDUCE GHG EMISSIONS

The overall efficiency of a power plant encompasses the efficiency of the various components of a particular generating unit. Sometimes these systems are unique to a generating unit, while in other instances these systems may be shared between generating units at a power plant site. This section will summarize the results of several U.S. and international studies which present options for improvements to power plant systems capable of increasing system heat rate efficiency and reducing GHG emissions. While the focus of this report is improving the efficiency of U.S. coal-fired power plants, studies from the international community are also presented as suggested improvements apply generally to coal-fired power plants. Further, a key report from an Asia-Pacific Economic Cooperation (APEC) working group report laid the groundwork for several U.S and international studies.

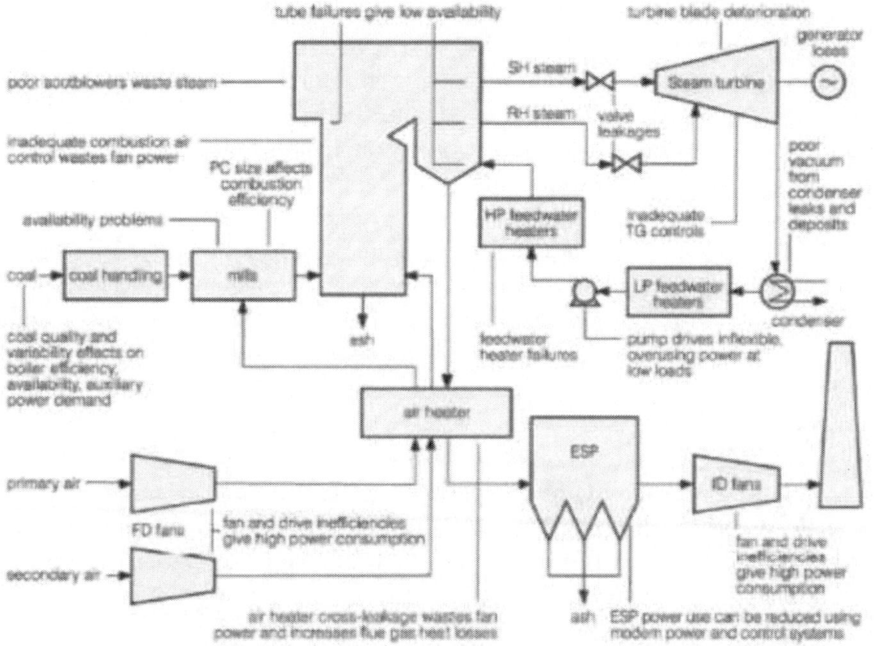

Source: International Energy Agency (IEA), Clean Coal Centre. Upgrading and Efficiency Improvement in Coal-fired Power Plants. Clean Coal Reports at http://www.iea-coal.org.uk/site/2010/publications-section/reports.

Notes: ID=induced draft fans are used to create a vacuum or negative air pressure in a system or stack; ESP= an electrostatic precipitator is a particulate collection device that removes particles from a flowing gas (such as air) using the force of an induced electrostatic charge ; FD=a forced draft fan is used to provide a positive pressure to a system ; LP=low pressure; HP=high pressure; PC=pulverized coal; RH=reheated ; SH=superheated; TG=turbine generator.

Figure 6. Areas of a Pulverized Coal Plant where Efficiency Loss Can Occur.

As coal-fired power plants age, they lose efficiency. Much of this loss in efficiency is due to mechanical wear on a variety of components resulting in heat losses, as can be seen in **Figure 6**. Lower power plant efficiency results in more CO2 being emitted per unit of electricity generated. The mode of operation (i.e., base load[26] vs. cyclical) also has a large effect on efficiency and fuel use.

The options most often considered for increasing the efficiency of CFPPs include equipment refurbishment, plant upgrades, and improved O&M schedules. Cost of the improvements is often compared to the expected return in increased efficiency as a primary determinant of whether to go forward with a program.

In 1999, the APEC region was responsible for 59% of the world's carbon dioxide emissions from fossil fuel combustion. Pulverized coal technologies accounted for 94% of coal-fired capacity in the region.[27] The 21 countries comprising APEC include Australia, Canada, China, Japan, the United States, and Russia.

According to an Asia-Pacific Economic Cooperation Working Group (APWG) study in 2001, projects to improve combustion, steam cycle, and O&M required low to medium costs, and these expenditures were predicted to produce as much as a 3.5% net overall efficiency improvement.[28] These improvements could also result in the largest overall reduction in CO_2 emissions of all the scenarios considered by the APWG study, since lower cost improvements were more likely to be adopted.

However, if reduction of carbon intensity is the goal (measured in grams of CO_2 emission per kiloWatt-hour of generation), the study found that switching of CFPPs to biomass ranked highest among the options considered (as biomass was considered carbon neutral), followed by fuel-switching to natural gas.

A subsequent APWG study in 2005[29] found that many older power plants in the Asia-Pacific region were operating well below their design efficiency. However, the study found that replacing the older CFPPs with new power plants was not practical because the expenditure for a new plant could not be justified by the improved performance.

Instead, efficiency and operational improvements were seen as a possible alternative considering a range of equipment upgrades and refurbishment options to various CFPP systems.

Table 1. Existing Coal-Fired EGU Efficiency Improvements Improvements Reported for Actual Improvement Projects

Efficiency Improvement Technology	Description	Reported Efficiency Increase See Note (a)
Combustion Control Optimization	Combustion controls adjust coal and air flow to optimize steam production for the steam turbine/generator set. However, combustion control for a coal-fired EGU is complex and impacts a number of important operating parameters including combustion efficiency, steam temperature, furnace slagging and fouling, and NOX formation. The technologies include instruments that measure carbon levels in ash, coal flow rates, air flow rates, CO levels, oxygen levels, slag deposits, and burner metrics as well as advanced coal nozzles and plasma assisted coal combustion.	0.15% to 0.84%
Cooling System Heat Loss Recovery	Recover a portion of the heat loss from the warm cooling water exiting the steam condenser prior to its circulation thorough a cooling tower or discharge to a water body. The identified technologies include replacing the cooling tower fill (heat transfer surface) and tuning the cooling tower and condenser.	0.2% to 1%
Flue Gas Heat Recovery	Flue gas exit temperature from the air preheater can range from 250 to 350°F depending on the acid dew point temperature of the flue gas, which is dependent on the concentration of vapor phase sulfuric acid and moisture. For power plants equipped with wet FGD systems, the flue gas is further cooled to approximately 125°F as it is sprayed with the FGD reagent slurry. However, it may be possible to recover some of this lost energy in the flue gas to preheat boiler feedwater via use of a condensing heat exchanger.	0.3% to 1.5%

Table 1. (Continued)

Efficiency Improvement Technology	Description	Reported Efficiency Increase See Note (a)
Low-Rank Coal Drying	Subbituminous and lignite coals contain relatively large amounts of moisture (15% to 40%) compared to bituminous coal (less than 10%). A significant amount of the heat released during combustion of low-rank coals is used to evaporate this moisture, rather than generate steam for the turbine. As a result, boiler efficiency is typically lower for plants burning low-rank coal. The technologies include using waste heat from the flue gas and/or cooling water systems to dry low-rank coal prior to combustion.	0.1% to 1.7%
Sootblower Optimization	Sootblowers intermittently inject high velocity jets of steam or air to clean coal ash deposits from boiler tube surfaces in order to maintain adequate heat transfer. Proper control of the timing and intensity of individual sootblowers is important to maintain steam temperature and boiler efficiency. The identified technologies include intelligent or neural-network sootblowing (i.e., sootblowing in response to real-time conditions in the boiler) and detonation sootblowing.	0.1% to 0.65%
Steam Turbine Design	There are recoverable energy losses that result from the mechanical design or physical condition of the steam turbine. For example, steam turbine manufacturers have improved the design of turbine blades and steam seals which can increase both efficiency and output (i.e., steam turbine dense pack technology).	0.84% to 2.6

Source: NETL. Available at Office of Air and Radiation, *Available and Emerging Technologies for Reducing Greenhouse Gas Emissions from Coal-Fired Electric Generating Units*, U.S. Environmental Protection Agency, October 2010, http://www.epa.gov/nsr/ghgdocs/electricgeneration.pdf.

Note; Reported efficiency improvement metrics adjusted to common basis by conversion methodology assuming individual component efficiencies for a reference plant as follows: 87% boiler efficiency, 40% turbine efficiency, 98% generator efficiency, and 6% auxiliary load. Based on these assumptions, the reference power plant has an overall efficiency of 32% and a net heat rate of 10,600 Btu/kWh. As a result, if a particular efficiency improvement method was reported to achieve a 1% increase in boiler efficiency, it would be converted to a 0.37% increase in overall efficiency. Likewise, a reported 100 Btu/kWh decrease in net heat rate would be converted to a 0.30% increase in overall efficiency.

National Energy Technology Laboratory Studies

The APWG results were amplified by a U.S. National Energy Technology Laboratory (NETL) study in 2008 which identified a list of potential methods to improve overall CFPP heat rate efficiency.[30,31] The results are shown in **Table 1**. The NETL study found that aside from a unit's age and steam cycle type, plant attributes such as location and emissions controls equipment did not account for the variations observed in plant efficiency. NETL then undertook an analysis of the efficiency of U.S. CFPPs, concluding that while the average efficiency was 32% in 2007, the efficiency of the top 10% was five points higher at 37.4%.[32] NETL suggested that if GHG emissions reduction was a goal, then heat rate efficiency improvements could enable a power plant to generate the same amount of electricity with lower CO_2 emissions.[33] In 2010, NETL completed a new study of U.S. CFPP efficiency,[34] dividing the results into 10 deciles of equal capacity (see **Table 2**). The generating units in the top 10% are diverse (i.e., they are not all new, large, supercritical plants), which NETL believes indicates an opportunity for overall fleet improvement. While on average the top decile consisted of units with larger capacities, higher steam pressures, higher load factors, and a higher percentage burning bituminous coal, there was significant overlap with the rest of the fleet. For example, the average steam pressure at the turbine of the top decile is around 3,000 psig, but there are ten units within the top decile with steam pressures in the 1,800 psig to 2,000 psig range. Also, the net nameplate capacity of the units in the top decile, while larger on average, ranged from 114 MW to 1,426 MW, indicating that even small plants can achieve higher than average efficiencies. The fleet generation-weighted average efficiency is 32.5%, while the top performing decile is over five percentage points higher (see **Table 3**). NETL therefore projected that average efficiency of the CFPP fleet could be raised from an average plant heat rate of 32.5% in 2008 to 36% overall.

Table 2. Segmentation Analysis of U.S. Coal-Fired Power Plants; Efficiency vs. Power Plant Type

Segment Criteria			Sub-Population Characteristics			Efficiency	
Unit Type	Coal Type	Size (MW)	Capacity (GW)	No. of Units	Generation (Billion, kWh)	Average	90th Percentile
Low Pressure Subcritical (600 – 1,600 psig)	Bit.	0 - 200	10.3	127	44	29.6%	33.0%
	Subbit.		4.6	59	26	27.5%	29.8%
	Other		0.6	7	2	27.4%	30.5%
High Pressure Subcritical (1,800 – 2,600 psig)	Bit.	0 – 200	21.6	134	112	32.1%	34.8%
		200 – 500	33.4	103	189	32.8%	35.6%
		500 +	29.7	48	176	32.7%	35.0%
	Subbit.	0 – 200	7.2	47	42	30.7%	32.5%
		200 – 500	31.2	97	191	31.4%	35.6%
		500 +	64.4	98	401	31.6%	33.8%
	Other		11.1	28	72	31.7%	35.1%
Supercritical (Over 3,334 psig)	Bit.		60.7	79	372	35.1%	37.3%
	Subbit.		15.0	20	90	35.2%	37.2%
	Other		8.1	13	55	31.8%	34.9%
Whole Fleet			298	860	1,772	32.5%	37.6%

Source: Figure 3, Segmentation Analysis of the CFPP Fleet. Improving the Efficiency of Coal-Fired Power Plants for Near Term Greenhouse Gas Emissions Reductions, DOE/NETL-2010/1411.

Notes: Bit= Bituminous; subbit. = subbituminous .

NETL sets forth a vision of 36% based on retirements of low efficiency units, and improvements within the best-in-class. Under a scenario where generation from coal is constant at the 2008 level, increasing the average efficiency from 32.5% to 36% reduces U.S. GHG by 175 MMmt/year or 2.5% of total U.S. GHG emissions in 2008.

According to NETL's analysis, retirements of lower efficiency units combined with increased generation from higher efficiency refurbished units, and advanced refurbishments with improved operation and maintenance, would be the key to potentially increasing average fleet efficiency beyond the best-in-class units.

The low pressure subcritical units and the 0-200 MW subbituminous units, in the sample studied, had 90th percentile efficiencies that were significantly lower than the rest of the fleet. Retiring these units and relying on increased generation from the more efficient segments to maintain constant coal generation yields a fleet efficiency target of 35.6%.

Table 3. Generation Weighted CFPP Efficiency By decile, 2008

Decile	Number of Units	Capacity (GW)	Capacity Factor	2008 Total Generation (Billion kWh)	2008 Generation-Weighted Efficiency (HHV)
1	194	30.5	62%	165	27.6%
2	102	30.3	67%	179	29.9%
3	88	30.7	65%	176	30.8%
4	86	30.6	69%	185	31.6%
5	75	30.7	70%	189	32.2%
6	83	30.8	66%	178	32.9%
7	71	31.0	68%	186	33.8%
8	79	30.6	68%	183	34.7%
9	61	30.8	67%	181	35.7%
10	53	30.7	74%	201	37.6%
OVERALL	892	307	69%	1,823	32.5%

Source: Improving Efficiency of Coal-fired Power Plants for Near Term CO_2 Reductions, NETL2.

NETL conceded that the fleet efficiency target of 36% did not consider installing scrubbers (to control emissions of sulfur oxides) at facilities without such controls.

> If efficiency upgrades were done in conjunction with installing sulfur scrubbers on 165-250 GW of the fleet, the efficiency target would be reduced 0.5 to 1 percentage points... NETL expects that the fleet achieving efficiencies approaching the best-in-class will require a combination of improved operational and maintenance practices, capital investments in unit refurbishments, and application of advanced technologies.[35]

NETL concluded its report on the opportunity to increase CFPP efficiency with the following suggestions on how to improve its analysis:

- Verification of coal generating unit efficiency data.
- Estimates of the cost of efficiency upgrades.
- Unit-specific data to enable estimation of the design heat rate for each generating unit.
- Case studies of efficiency upgrades at generating units, including modeling, to provide concrete examples of the opportunity to improve efficiency.
- Detailed analyses of scenarios where some of the generating units are retired, some are refurbished and up-rated, and others are fixed with sulfur, oxides of nitrogen, and mercury controls.
- Analyses of how more efficient coal plants will dispatch, what other generating sources will be displaced, and the overall effect on GHG emissions.

Environmental Protection Agency Study

The U.S. Environmental Protection Agency (EPA) issued a report[36] in 2010 which affirmed that there was a direct relationship between CFPP efficiency and CO_2 emissions. EPA recognized that the level of CO_2 emissions potentially released from a given coal-fired electric generating unit (EGU) (i.e., power plant) depends on the type of coal burned, the overall efficiency of the power generation process, and use of air pollution control devices.

> In addition to the lower CO_2 emissions rate per unit of heat input (lbs CO_2/MMBtu), due to the inherent moisture in subbituminous and lignite coals, all else being equal a bituminous coal-fired boiler is more efficient than a corresponding boiler burning subbituminous or lignite coal. Therefore, switching from a low to a high-rank coal will tend to lower GHG emissions from the utility stack.[37]

As the thermal efficiency of a coal-fired power plant is increased, less coal is burned per kWh of electricity generated, and there is a corresponding decrease in CO_2 and other air emissions. The greater the output of electric energy for a given amount of fuel energy input, the higher the efficiency for the electric generation process. Heat rate is another common way to express efficiency. Power plants that are more efficient typically have lower heat rates.[38]

Although EPA states there is a direct relationship between coal-fired EGU efficiency and CO_2 emissions, EPA recognized that other factors must be considered when comparing the effectiveness of GHG control technologies to improve the efficiency of a given coal-fired EGU. The study states that the actual overall efficiency that a given coal-fired EGU achieves is determined by the interaction of a combination of site-specific factors that impact efficiency to varying degrees, including:

> EGU thermodynamic cycle – EGU efficiency can be significantly improved by using a supercritical or ultra-supercritical steam cycle.
>
> EGU coal rank and quality – EGUs burning higher quality coals (e.g., bituminous) tend to be more efficient than EGUs burning lower quality coals (e.g., lignite).
>
> EGU plant size – The electric-generating capacity of EGUs ranges from approximately 25 to 1,300 MW. Assuming an EGU efficiency of 33% (a typical efficiency for existing coal-fired EGUs), this corresponds to a heat input range of 250 to 13,400 MMBtu/hr.
>
> EGU efficiency generally increases with size because the boiler and steam turbine losses are lower for larger equipment. However, as equipment size increases [beyond a certain point] the differences in these losses start to taper off.
>
> EGU pollution control systems – The electric power consumed by air pollution control equipment reduces the overall efficiency of the EGU.
>
> EGU operating and maintenance practices – The specific practices used by an individual electric utility company [including] combustion optimization, equipment maintenance, can affect EGU efficiency.
>
> EGU cooling system – The temperature of the cooling water entering the condenser can have impacts on steam turbine performance. Once-through cooling systems can have an efficiency advantage over recirculating cooling systems (e.g., cooling towers). However, once-though cooling systems typically have larger water related ecological concerns than recirculating cooling systems.
>
> EGU geographic location – The elevation and seasonal ambient temperatures at the facility site ... may have a measureable impact on EGU efficiency. At higher elevations, air pressure is lower and less oxygen is available for combustion per unit volume of ambient air than at lower elevations. Cooler ambient temperatures theoretically could increase the overall EGU efficiency by increasing the draft pressure of the boiler flue gases and the condenser vacuum, and by increasing the efficiency of a condenser recirculating cooling system.
>
> EGU load generation flexibility requirements – Operating an EGU as a base load unit is more efficient than operating an EGU as a load cycling unit to respond to fluctuations in customer electricity demand.

EGU equipment manufacturers – The efficiency specifications of major EGU components such as boilers, turbines, and electrical generators provided by equipment manufacturers can affect EGU efficiency.

EGU plant components – EGUs using the optimum number of feedwater heaters, high efficiency electric motors, variable speed drives, better materials for heat exchangers, etc. tend to be more efficient.[39]

Based on the factors above, EPA concludes that coal-fired power plants "identical in design but operated by different utility companies in different locations may have different efficiencies. Thus, the level of effectiveness of a given GHG control technology used to improve the efficiency at one coal-fired EGU facility may not necessarily directly transfer to a coal-fired EGU facility at a different location."[40]

EPA recognized in the study that a number of technologies to improve power plant efficiency are available for application to existing coal-fired EGU projects which can incrementally improve thermal overall efficiency, and expanded on the efficiency improvements reported by NETL in its 2008 report on efficiency improvement projects (shown in **Table 1**).

International Energy Agency Study

The International Energy Agency released a study[41] in 2013 looking at opportunities to reduce CO_2 emissions using upgrades and efficiency improvements at CFPPs. IEA concluded that substantial improvements (i.e., retrofits) may be seen as cost-effective if these economically restore the efficiency of a power plant.

> Despite involving substantial outlay (typically US$100–200 million), retrofits will provide a payback in restored generation, fuel saving, extended plant life, and, in some countries, CO_2 emissions cost savings. There are also benefits of reduced specific emissions of other pollutants.[42]

Retrofits include turbine upgrades, condenser optimization, increasing the capacity and efficiency of air-cooled condensers, boiler system improvements, and improvements to other systems where energy losses can occur (as shown in **Figure 6**).

However, IEA affirmed that major plant retrofits and upgrades (i.e., conversion of subcritical PC units to super- or ultra-supercritical PC units) would raise efficiencies more substantially. IEA used the example of a conversion of a subcritical 500 MW unit in the United Kingdom to a supercritical pressure, which was projected to raise net generation efficiency from 38% to 44% (on a lower heating value basis). The upgrade was projected to reduce CO_2 emissions by 500,000 tonnes per year.

USING RENEWABLES TO IMPROVE COAL PLANT EFFICIENCY

Heat rate improvement could potentially be achieved using renewable technologies to either provide heat to reduce heat losses at various points in the steam cycle, or to provide power to the equipment used to curb these heat losses, thus curbing on-site equipment electricity use. One such hybrid coal-solar power plant is already in operation in the United States, at the Xcel Cameo Generating Station in Colorado.

> The demonstration project is expected to cut the use of coal at the power plant by around two or three percent, and could be scaled up to cut it by 10 percent. The system works through a series of parabolic trough solar collectors made of glass mirrors. On sunny days the mirrors concentrate the solar radiation onto a line of receiver tubes filled with a heat transfer fluid (mineral oil). The solar energy heats the circulating oil to about 300°C (575°F). The heated oil is then fed to a heat exchanger where the heat is transferred to water to heat it to around 200°C (407°F) before it enters the boiler. Having hotter water entering the boiler means less coal is needed to heat it and produce the steam that turns the turbine to generate electricity.[43]

Alternatively, using biomass has been suggested to co-fire with coal in a CFPP, or to replace coal altogether.

> Combining the use of biomass with coal can be beneficial, particularly from an environmental standpoint although any such process may have its limitations or drawbacks. Each coal type and biomass feedstock has different characteristics although by combining the two, it may be possible to capitalize on the advantages of each, and minimize their individual disadvantages. An effective way is via [gasification and production of syngas, a mixture of hydrogen and carbon monoxide], and useful operating experience has been achieved in a number of large-scale coal-fuelled gasification and IGCC plants ... It also has the potential to

> form the basis of systems that combine coal and biomass use with other renewable energy technologies to create clean, efficient energy-production systems. Thus, various hybrid energy concepts, some based on coal/biomass [gasification], have been proposed or are in the process of being developed or trialed. Some propose to add yet another element of renewable energy to the system, generally by incorporating electricity generated by intermittent renewables such as wind or solar power. A number also aim to incorporate some form of carbon capture and storage.[44]

As biomass is generally considered carbon neutral,[45] co-firing coal with biomass can provide advantages for electric power generation. However, using biomass on a large, commercial scale has a number of potential issues. Since the heating value and bulk density of biomass is lower than coal, the necessary volumes to be harvested and handled can be substantial, and the type and availability of different biomass materials tends to vary considerably with location. A potential source of biomass in the United States could be wastes from the forest products industry.[46]

POTENTIAL BARRIERS TO IMPLEMENTING EFFICIENCY UPGRADES

Government regulations, regulatory regimes (i.e., competitive markets or traditional cost of service regulation), and industry factors (such as growth or a lack of growth in demand) all affect the motivation for implementing improvement projects at coal-fired power plants. These external forces add to the internal evaluation of the costs vs. benefits of improvements for a particular unit. This section will look at a few of these external forces which may hamper the implementation efficiency projects.

New Source Review

The New Source Review program was designed to prevent the degradation of air quality from the construction of new facilities or modification of existing facilities which have potentially harmful emissions. NSR was established by Congress as part of the 1977 Clean Air Act Amendments (P.L. 95-95).

The NSR process requires power plant operators to undergo a review for environmental controls if they build a new power generating unit, and to impose the Best Available Control Technology, as defined by the state permitting authority (or in some cases EPA). Efficiency improvements to power plants that reduce regulated pollutants should not theoretically trigger NSR requirements, unless the improvements result in an increase in emissions (e.g., because the modified plant operates for more hours). Establishment of a pre-improvement emissions baseline before and a post-improvement emissions report after efficiency upgrades seems like a logical step, but may not be easily achievable on a consistent basis.[47] There are also ambiguities in the law which may serve to hamper efficiency projects from going forward.

> Power plants built prior to 1971 are exempted from the limits on criteria pollutant emissions contained in the Clean Air Act, but may lose that exemption and be forced to undergo an NSR if the EPA determines that the plant has undergone non-routine maintenance which increase emissions.
>
> The power generation industry widely views the NSR process as an obstacle to power plant efficiency improvement projects. In a 2002 report to the President, the Environmental Protection Agency concurs, stating "that NSR as applied to existing plants discourages projects that would have provided needed capacity or efficiency improvements."[48]

According to NETL, there are two critical issues with respect to the NSR and efficiency improvements:[49]

- The definition of "routine maintenance, repair, and replacement" projects, and
- Whether a unit's emission rate or its total emissions over a specified time should be used to determine if an efficiency improvement project increases emissions.

NETL further stated that several developments since the passing of the 1977 Clean Air Act Amendments have failed to clarify these issues and made efficiency improvements risky and less appealing to plant operators. On the other hand, NETL also stated that "[u]nder future scenarios, the added value of higher efficiency in meeting GHG emissions limits, may prevent the NSR from being utilized as a barrier to capital investments aimed at improving power plant efficiency."[50]

Fuel Prices for Electricity and Regulatory Uncertainty

Increases in the domestic production of natural gas (primarily due to hydraulic fracturing of gas shales) are causing a dramatic change in electric power production decisions. A recent decline in natural gas prices has come with increased production of natural gas, and the resulting decline in coal consumption for power generation. In April 2012, for the first time in U.S. history, the amount of electricity generation from natural gas equaled that of coal, according to EIA statistics, with each representing about 32% of the market.

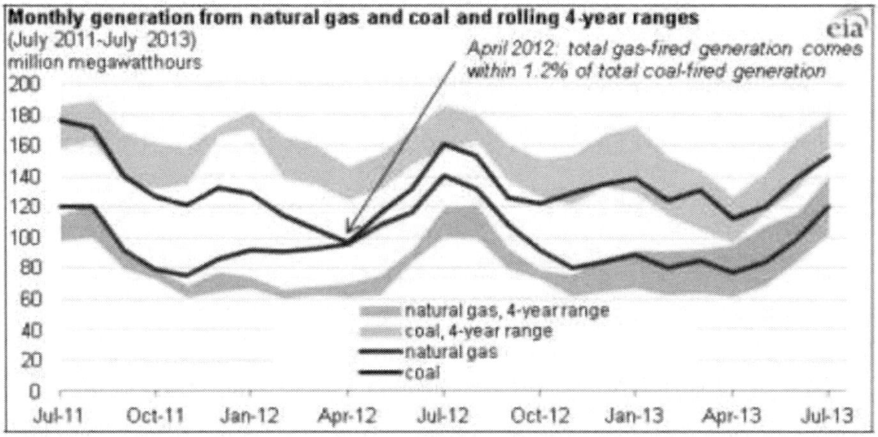

Source: U.S. Energy Information Administration, Electric Power Monthly.

Notes: The rolling four-year range for each fuel source represents the maximum and minimum amount of each fuel source that is consumed for electricity generation for the same month during the previous four years.

Figure 7. Recent Electric Power Generation Trends; July 2011 to July 2013.

However, as **Figure 7** illustrates, coal has recently rebounded and regained market share for electricity generation as natural gas prices have risen. EIA states that since May 2013, "a combination of higher prices for natural gas and increased demand for electricity during the summer months led electric systems across much of the country to increase their use of coal-fired units."[51]

Growth in demand for electricity is expected to be low over the coming decades.[52] By itself, this would ordinarily be expected to reduce new power plant construction, and might prompt some coal plant owners to consider

projects to increase efficiency from existing plants. However, lower natural gas prices, higher coal prices, slower economic growth, and the implementation of new environmental rules[53] are all expected to play a role in the level of coal plant retirement decisions.[54]

POSSIBLE APPROACHES TO ENCOURAGE EFFICIENCY IMPROVEMENT

As previously mentioned, EPA has been directed by the President to propose guidelines for GHG emissions reduction from existing coal-fired power plants. The general assumption is that EPA will establish some target for emissions on a state or plant-by-plant basis, with companies free to decide how they will achieve the reduction (i.e., with efficiency, fuel switching, retirement of older units, priority dispatch for cleaner units, etc.), and with emissions averaging, banking, and trading of emissions credits playing a role. Within such a system, efficiency improvements can be an important contributor.

While the details of the proposal are unknown at this time, the proposal may take into account a number of factors (such as the remaining useful life of the existing source), and could be less stringent than the proposal for new sources of power plants emissions of GHGs.[55]

Power plant efficiency may be another factor which EPA may potentially consider in its guidelines for existing sources. NETL observed in its 2010 report[56] that based on a scenario where CFPP generation was constant at the 2008 level, increasing the average efficiency from 32.5% to 36% could reduce U.S. GHG emissions by 175 MMmt per year or 2.5% of total U.S. emissions in 2008.[57] NETL conceded that barriers existed to achieving a higher average fleet efficiency level, citing the power generation industry's focus on availability (focused on the profitability of coal-fired generating units), inconsistent cost pass through possibilities (some deregulated areas have cost pass through clauses, and zero or negative incentives in many areas for reduced fuel use), fear of triggering New Source Review, and uncertainty about GHG regulations (which could lead to very short payback periods for improvements).[58]

If power plant efficiency is an option EPA proposes for state consideration, the question then may be asked how a fleet-wide improvement program could be achieved in United States. One possible approach might be to follow NETL's suggestion of using the top decile of CFPP efficiency as a benchmark for U.S. fleet efficiency, used with an efficiency frontier.[59] Using statistical methods, benchmarks[60] could be used to improve efficiency of the CFPP fleet.[61]

NETL observes that while some improvements could be "relatively inexpensive" (for example, improved O&M, more frequent or pro-active maintenance), other improvements could be "very expensive" (for example, improvements bundled with a new SO_2 scrubber, or turbine overhauls or heat exchanger replacement). But NETL notes that "if each plant achieved their maximum efficiency each year, 5% reduction in CFPP carbon dioxide emissions" could result.[62]

According to NETL's analysis, retirements of lower efficiency units combined with increased generation from higher efficiency refurbished units, and advanced refurbishments with improved operation and maintenance, would be the key to increasing average fleet efficiency.

Efficiency improvements could be incentivized using an efficiency frontier. The selection of appropriate incentives would then encourage CFPP owners to undertake improvements or retire lower efficiency units. Such incentives could include possible tax rate reductions for CFPP owners matched in some manner to the cost of the improvements, or accelerated book depreciation[63] rates for cost recovery. Penalties could perhaps be used to encourage prompt retirement of the lowest efficiency units.

The incentives could be in place over a defined period of years, with incentives reduced during the period to encourage action sooner rather than later.

The effective period of incentives would have to be sufficiently long enough to allow equipment orders to be satisfied, and simultaneous work to progress at multiple CFPP sites across the country which might seek to make improvements considering equipment lead times and workforce availability. The benchmark could be revised periodically. This could allow newer power plants to live out a service life matched to the most efficient operation achievable for a particular type of CFPP, based on industry statistics. This could allow companies that have made substantial investments in pollution controls an opportunity to recover these investments.

However, EPA's expected proposal on standards for GHG emissions from existing coal-fired power plants will be a primary factor in determining whether efficiency improvements will be cost effective in the near term.

CONCLUSION AND POLICY OPTIONS

The efficiency of coal-fired power plants decreases over time as components and systems degrade with age and use. Good O&M practices can slow down the loss of efficiency, but older power plants will not be as efficient as newer plants with more technologically advanced and newer systems. But simply replacing old power plants with newer plants is rarely cost effective as the relative increase in power output seldom justifies the cost. CFPPs that are more efficient emit less CO_2 per unit of electricity produced because they use less coal. Making improvements to increase the efficiency of CFPPs (while producing the same electrical output) could result in a significant reduction in CO_2 emissions. According to several of the studies summarized in this report, the major improvements in GHG emissions would likely result from major retrofits in technology, or conversions to natural gas (or possibly biomass) as a fuel.

Detailed information on the actual cost of efficiency-enhancing improvements is not readily available, as concerns over confidentiality and competitiveness with regard to actual projects has largely prevented the sharing of such information. The studies referenced in this report largely show relative information on cost (i.e., high, medium, low), estimate the cost effectiveness of improvements, or mention general cost levels. The case studies have reported costs of efficiency improvements and actual increases in efficiency for specific power plant.[64] But these are considered as useful for estimating a range of costs for improvements rather than actual guides for costs, since each power plant has its own design characteristics and maintenance history. Actual cost information would require a technical evaluation, and a cost vs. benefit analysis to obtain reliable cost estimates for the options under consideration, taking into account site-specific conditions.

Other potential roles exist for Congress. For example, legislation could use tax incentives to encourage energy efficient upgrades for CFPPs which were placed in-service after a certain date. This would allow newer units with environmental controls to recover the cost of scrubbers and other systems, and allow "newer" units to continue operations over a "reasonable" service life, and recover the cost of environmental improvements.

Another approach might be to use a federal energy efficiency standard to accomplish a similar goal as an efficiency frontier, but allow the states to design the program based on local fuel resources, the age of power plants under their jurisdiction, and other criteria defined in legislation. The efficiency standards could increase over time, and require CFPPs not meeting these standards to retire.

Deference to state authorities and regional compliance strategies have been suggested by observers with regard to EPA's deliberations over GHG reduction for existing CFPPs. State public utilities commissions (or similar entities) often require utilities to conduct book depreciation studies (either in connection with rate cases or independent of rate cases). Such studies commonly examine the physical condition of power plants, and the utility's recovery of its investment in electric plant. Federal legislation could tie incentives for efficiency improvements to such studies and direct states to meet individual or regional goals for GHG reduction.

End Notes

[1] Greenhouse gases, according to EPA, are any gas that absorbs infrared radiation in the atmosphere. There are six greenhouse gases addressed by EPA regulatory actions: carbon dioxide (CO2), methane (CH4), nitrous oxide (N2O), and fluorinated gases—sulfur hexafluoride (SF6), hydrofluorocarbons (HFCs), and perfluorocarbons (PFCs). Carbon dioxide is the most prevalent GHG produced by combustion of fossil fuels. See http://www.epa.gov/climatechange/ ghgemissions/gases.html.

[2] EPA's proposed standard for new power plants would require new coal units to achieve the emissions of a natural gas combined-cycle unit, which would likely require coal units to employ CCS. Environmental Protection Agency, *2013 Proposed Carbon Pollution Standard for New Power Plants*, September 23, 2013, http://www2.epa.gov/carbon-pollution-standards/2013-proposed-carbon-pollution-standard-new-power-plants.

[3] "The President directed EPA to propose such guidelines by June 2014 and to finalize them a year later. Using these guidelines, states will be required to develop performance standards for existing sources. These could be less stringent than the NSPS—taking into account, among other factors, the remaining useful life of the existing source." For a discussion of current EPA regulations and coal-fired power plant GHG emissions, see CRS Report R43127, *EPA Standards for Greenhouse Gas Emissions from Power Plants: Many Questions, Some Answers*, by James E. McCarthy.

[4] Congress established the New Source Review permitting program as part of the 1977 Clean Air Act Amendments. NSR is a preconstruction permitting program that serves two important purposes. First, it ensures that air quality is not significantly degraded from the addition of new and modified factories, industrial boilers and power plants. In areas with unhealthy air, NSR assures that new emissions do not slow progress toward cleaner air. In areas with clean air, especially pristine areas like national parks, NSR assures that new emissions do not significantly worsen air quality. Second, the NSR program assures people that any large new or modified industrial source in their neighborhoods will be as clean as possible, and that advances in pollution control occur concurrently with industrial expansion. NSR permits are legal documents that the facility owners/operators must abide by. The permit

specifies what construction is allowed, what emission limits must be met, and often how the emissions source must be operated.

See Environmental Protection Agency, *New Source Review*, June 11, p. 2013, http://www.epa.gov/NSR/.

[5] See CRS Report R41325, Carbon Capture: A Technology Assessment.

[6] A simple cycle natural gas power plant burns natural gas in a combustion turbine (like a jet engine) to turn a generator which produces electricity. When used in a combined cycle mode, the hot exhaust gases from the combustion turbine are used to generate steam to turn a steam turbine to efficiently create additional electricity. Nuclear fuel does not "combust," but it does produce heat to make steam.

[7] See 2012 Form EIA-860 Data - Schedule 3, 'Generator Data' (Operable Units Only).

[8] See Electric Generating Capacity, U.S. Energy Information Administration's Annual Energy Outlook at http://www.eia.gov/forecasts/aeo/er/early_ elecgen.cfm.

[9] The heating value of any fuel is the energy released per unit mass or per unit volume of the fuel when the fuel is completely burned. Higher heat value fuels liberate more energy per unit of mass or volume.

[10] "To optimize overall efficiency for a given [power plant or electric generating unit] EGU, the unit is operated under conditions such that nearly all of the fuel carbon is converted to CO2 during the combustion process. Methane is emitted during the mining and transport of coal but is not a significant by-product of EGU coal combustion. Fluorinated gases are not formed by coal combustion. Sulfur hexafluoride might be used at the power plant switchyard, but the switchyard is not typically considered part of the EGU." Sector Policies and Programs Division, *Available and Emerging Technologies for Reducing Greenhouse Gas Emissions from Coal-Fired Electric Generating Units*, U.S. Environmental Protection Agency, Office of Air Quality Planning and Standards, October 2010, http://www.epa.gov/ nsr/ghgdocs/electricgeneration.pdf.

[11] The thermodynamic cycle that is an ideal standard for comparing performance of heat-engines, steam power plants, steam turbines, and heat pump systems that use a condensable vapor as the working fluid. See http://www.eia.gov/ tools/glossary/index.cfm.

[12] Subcritical steam operating conditions are generally at pressures of 2,400 pounds per square inch gage (psig) [i.e., relative to atmospheric pressure] per 1,000 [degrees Fahrenheit] °F of superheated steam. Supercritical steam cycles typically operate at 3,600 psig, with 1,000 °F to 1,050 °F steam conditions. See American Electric Power Company, *Pulverized Coal Technologies*, 2013, http://www.aep.com/about/IssuesAnd Positions/Generation /Technologies/ PulverizedCoal.aspx.

[13] Cyclone-fired boilers are used for coals with a low ash fusion temperature which are difficult to use with pulverized coal combustion (PCC). 80-90% of the ash leaves the bottom of the boiler as a molten slag, thus reducing the load of fly ash passing through the heat transfer sections to the precipitator or fabric filter to just 10-20% of that present. As with PCC, units operate at close to atmospheric pressure, simplifying the passage of coal and air through the plant. Steam is generated in heat transfer tubes, driving a steam turbine and generator. See http://www.iea-coal.org.uk/site/ 2010/database-section/ccts/cyclone-fired-wet-bottom-boilers?

[14] A method of burning particulate fuel, such as coal, in which the amount of air required for combustion far exceeds that found in conventional burners. The fuel particles are continually fed into a bed of mineral ash in the proportions of one part fuel to 200 parts ash, while a flow of air passes up through the bed, causing it to act like a turbulent fluid. See http://www.eia.gov/tools/glossary/index.cfm.

[15] Coal, water, and oxygen are fed to a gasifier, which produces syngas. This medium-Btu gas is cleaned (particulates and sulfur compounds removed) and is fed to a gas turbine. The hot exhaust of the gas turbine and heat recovered from the gasification process are routed through a heat-recovery generator to produce steam, which drives a steam turbine to produce electricity. See http://www.eia.gov/tools/glossary/index.cfm.

[16] International Energy Agency, *Upgrading and efficiency improvement in coal-fired power plants*, No. 13/9, August 2013, http://www.iea-coal.org.uk/documents/83185/8784/Upgrading-and-efficiency-improvement-in-coal-fired-powerplants,-CCC/221.

[17] CRS Report R42950, *Prospects for Coal in Electric Power and Industry*, by Richard J. Campbell, Peter Folger, and Phillip Brown.

[18] Efficiency can be calculated using the higher heating value (HHV) or the lower heating value (LHV) determined for the fuel. The HHV is the heating value directly determined by calorimetric measurement of the fuel in the laboratory. The LHV is calculated using a formula to account for the moisture in the fuel (i.e., subtract the energy required to vaporize the water in the coal and thus not available to produce steam) that is a smaller value than the HHV. Consequently, the HHV efficiency for a given EGU is always lower than the corresponding LHV efficiency, because the reported heat input is larger for the same output.

[19] Therefore, 10,444 BTU of energy from coal produces 3,412 BTU per each kiloWatt-hour of electricity. See EIA, What is the efficiency of different types of power plants? at http://www.eia.gov/tools/faqs/ faq.cfm?id= 107&t=3. Coal includes anthracite, bituminous, subbituminous and lignite coal. Waste coal and synthetic coal are also included.

[20] Joshua Linn, Erin Mastrangelo, and Dallas Burtraw, *Regulating Greenhouse Gases from Coal Power Plants under the Clean Air Act*, Resources for the Future, RFF DP 13-05, February 2013, http://www.rff.org/RFF/ documents/RFFDP-13-05.pdf.

[21] János Beér, *High Efficiency Electric Power Generation; The Environmental Role*, Massachusetts Institute of Technology, October 17, 2006, http://mitei.mit.edu/system/files/beer-combustion.pdf. (BEF).

[22] Deborah Adams, *Efficiency Upgrades for Existing Coal-fired Power Plants*, International Energy Agency—Clean Coal Centre, Presentation, Gliwice, Poland, September 24, 2009, p. 8, http://cleancoal.polsl.pl/pdf/Adams.pdf.

[23] Ibid.

[24] George Booras and Neville Holt, *Pulverized Coal and IGCC Plant Cost and Performance*, Electric Power Research Institute, Gasification Technologies 2004, October 2004, http://citeseerx.ist.psu.edu/ viewdoc/download?doi= 10.1.1.168.2103&rep=rep1&type=pdf.

[25] International Energy Agency, Bulletin No. 13/9, *Upgrading and efficiency improvement in coal-fired power plants*, August 2013, http://www.iea-coal.org.uk/documents/83185/8784/Upgrading-and-efficiency-improvement-in-coalfired-power-plants,-CCC/221.

[26] Base load plant: A plant, usually housing high-efficiency steam-electric units, which is normally operated to take all or part of the minimum load of a system, and which consequently produces electricity at an essentially constant rate and runs continuously. These units are operated to maximize system mechanical and thermal efficiency and minimize system operating costs. See http://www.eia.gov/tools/glossary/index.cfm.

[27] See Table 3-7, Distribution of Existing Capacity by Fuel and Type of Energy Technology for APEC Economies as of November, 2000 at http://www.egcfe.ewg.apec.org/projects/CO2_Phase1_Study_2000.pdf.

[28] See Scenario E-4, Options to reduce CO2 Emissions from Electricity Generation in the APEC Region. Energy Working Group Project EWG 04/2000.

[29] Asia-Pacific Economic Cooperation Working Group, Costs and Effectiveness of Upgrading and Refurbishing Older Coal-Fired Power Plants in Developing APEC Economies, Energy Working Group Project EWG 04/2003T, June 2005, http://www.egcfe.ewg.apec.org /projects/ UpgradePP_Report_2005.pdf.

[30] DOE—National Energy Technology Laboratory, *Reducing CO2 Emissions by Improving the Efficiency of the Existing Coal-fired Power Plant Fleet*, DOE/NETL-2008/1329, July 2008, http://www.netl.doe.gov/energy-analyses/ pubs/CFPP%20Efficiency-FINAL.pdf. (NETL)

[31] It is important to note that improved efficiency in one area of a power plant can also help improve efficiency in other areas. "... [O]ptimization of the combustion process can give valuable benefits in efficiency and costs. The gain may typically be about 0.1% to 0.15% in fuel cost saving, efficiency and CO2 emissions ...Improvements in combustion efficiency can be achieved in parallel with other improvements, for example, reductions in primary NO_x production from replacement burners and new air supply arrangements." See Colin Henderson, Upgrading and Efficiency Improvement in Coal-fired Power Plants, International Energy Agency Clean Coal Centre, CCC/221, August 2013, http://www.iea-coal.org.uk/site/2010/publications-section/reports.

[32] See Table 2, Reducing CO2 Emissions by Improving the Efficiency of the Existing Coal-fired Power Plant Fleet. DOE/NETL-2008/1329.

[33] While all the listed areas for improvement could be targets, the actual improvements considered would depend on an evaluation of site-specific conditions.

[34] Phil DiPietro, Katrina Krulla, DOE/National Energy Technology Laboratory, *Improving the Efficiency of Coal-fired Power Plants for Near Term Greenhouse Gas Emissions Reductions*, DOE/NETL-2010/1411, April 16, 2010, http://www.netl.doe.gov/energy-analyses/pubs/ ImpCFPPGHGRdctns_0410.pdf. (NETL2).

[35] Ibid.

[36] Office of Air and Radiation, *Available and Emerging Technologies for Reducing Greenhouse Gas Emissions from Coal-Fired Electric Generating Units*, U.S. Environmental Protection Agency, October 2010, http://www.epa.gov/nsr/ ghgdocs/electricgeneration.pdf. (EPAEff).

[37] Ibid.

[38] Bituminous coals thus tend to have a lower HHV efficiency than LHV efficiency, and higher moisture content subbituminous coals and lignite have HHV efficiency approximately 3 to 5 percentage points lower than the corresponding LHV efficiency.

[39] EPAEff.

[40] EPAEff.

[41] *Upgrading and Efficiency Improvement in Coal-fired Power Plants*. Clean Coal Reports at http://www.ieacoal.org.uk/site/2010/publications-section/reports.

[42] Ibid.

[43] Lin Edwards , *First ever hybrid solar-coal power plant operating* , PhysOrg.com, July 12, 2010, http://phys.org/ news198135102.html.

[44] Stephen Smith, *Combining Renewable Energy with Coal*, IEA Clean Coal Centre, CCC/223, September 2013, http://www.iea-coal.org.uk/site/2010/ publications-section/reports,CCC /223.

[45] There are challenges to EPA's decision to treat biomass as carbon neutral. See CRS Report R41603, *Is Biopower Carbon Neutral?*, by Kelsi Bracmort.

[46] CRS Report R40565, Biomass Resources: The Southeastern United States and the Renewable Electricity Standard Debate, by Richard J. Campbell.

[47] International Energy Agency / Coal Industry Advisory Board, *Power Generation from Coal - Measuring and Reporting Efficiency Performance and CO2 Emissions*, 2010, http://www.iea.org/ciab/papers/ power_generation_from_coal.pdf.

[48] NETL, page 4.

[49] Ibid.

[50] Ibid.

[51] See EIA, Today in Energy, at http://www.eia.gov/todayinenergy/ detail.cfm?id=11391.

[52] The long-run trend of slowing growth in electricity use relative to economic growth in the United States is expected to continue, with the rate of projected growth in electricity use at less than half the rate of economic growth. See http://www.eia.gov/todayinenergy /detail.cfm?id=1049.

[53] The EPA is obligated under Section 111(d) of the Clean Air Act to promulgate guidelines for GHG emission standards for existing coal-fired power plants. EPA has been directed by the President to propose such guidelines by June 2014, and to finalize them a year later.

[54] Energy Information Adminstration, *Projected retirements of coal-fired power plants*, July 31, 2012, http://www.eia.gov/todayinenergy/ detail.cfm?id=7330.

[55] Op. cit. CRS R43127.

[56] Op. cit. NETL.

[57] NETL suggested that the cost of such improvements could be a point for further study.

[58] "With many different methods used to express efficiency performance, it is often difficult to compare plants, even before accounting for any fixed constraints such as coal quality and cooling-water temperature." See *Improving Efficiency of Coal-fired Power Plants for Near Term CO2Reductions*, National Energy Technology Laboratory, Presentation, 2009 at http://www.netl.doe.gov/energy-analyses/pubs/Impr_Effcy_of_CFPP_CO2_Redctns_1109. pdf. (NETL3).

[59] According to Cambridge Economic Policy Associates, the "Efficiency Frontier" essentially relates technical efficiency (the ability to produce the maximum level of outputs from a given set of inputs) to allocative efficiency (reflecting the extent to which the inputs are optimal for a given set of prices using a given technology). These two measures are then combined to form a measure of total economic efficiency or total cost efficiency (not thermal efficiency), which represents a notion of "best practices" at a given point in time. By extension, those firms operating at or below the efficiency frontier are "efficient," and those firms operating above the frontier are "inefficient." See http://www.ppparbiter.org.uk/ pdf_folder/cepa_r1ax1_0703.pdf.

[60] Benchmarking is the process of comparing an individual organization's business processes and performance metrics to best practices in the same industry or applicable best practices from other industries.

[61] NETL3, and B.F. Roberts and Lessly Goudarzi, *Efficient Heat Rate Benchmarks for Coal-Fired Generating Units*, 1998, http://www. onlocationinc.com/heatratepaper.pdf.

[62] NETL3.

[63] Book Depreciation is a regulatory accounting concept which involves the allocation of the cost of an asset over its expected useful service life in a manner that systematically charges the cost of the asset over the period of time it is in service. Book depreciation may be charged at a faster or slower rate than allowed by the Internal Revenue Service, in order to provide management with a realistic view of the gradually diminishing value of the company's assets.

[64] As an example of what can be achieved, a 404 MW coal-fired plant reported a 4,523 kW plant capacity increase (1.1%) and 0.7% point improvement in plant efficiency from the following low-cost steam turbine upgrades:

Replaced radial spiral strips on high-pressure (HP) and intermediate pressure (IP) sections.

Installed retractable interstage packing.

Repaired solid particle erosion (SPE) damage to stationary diaphragms.

Repaired station diaphragm flow path SPE in IP section. APEC, page 72.

In: Improving the Efficiency ...
Editor: Roland Ludovic
ISBN: 978-1-63117-590-9

Chapter 2

IMPROVING THE EFFICIENCY OF COAL-FIRED POWER PLANTS FOR NEAR TERM GREENHOUSE GAS EMISSIONS REDUCTIONS*

Phil DiPietro and Katrina Krulla

EXECUTIVE SUMMARY

In 2008, the U.S. coal-fired power plant (CFPP) fleet had a generation-weighted average efficiency of 32.5% while the top ten percent of the fleet had an efficiency of 37.6%, five percentage points higher. The generating units in the top ten percent are diverse (they are not all new, large, super critical plants), indicating an opportunity for fleet-wide efficiency improvement. The National Energy Technology Laboratory (NETL) segmented the fleet into 13 groups based on characteristics that limit efficiency, and calculated the best-in-class efficiency within each group. Based on each group achieving an average efficiency equal to its 90th percentile, the overall CFPP fleet average efficiency would be 35.2%. NETL sets forth a vision of 36% based on retirements of low efficiency units, and improvements within the best-in-class. Under a scenario where generation from coal is constant at the 2008 level, increasing the

* This is an edited, reformatted and augmented version of a National Energy Technology Laboratory report, DOE/NETL-2010/1411, dated April 16, 2010.

average efficiency from 32.5% to 36% reduces U.S. GHG by 175 MMmt/year [1] or 2.5% of total U.S. GHG emissions in 2008.

Decile Analysis

NETL examined coal generating unit efficiency data from the Energy Velocity (EV) database[1], [2] for the years 1998 through 2008. Figure 1 shows the performance of the CFPP fleet in 2008 ranked by efficiency and broken into ten deciles of equal capacity. The fleet generation-weighted average efficiency is 32.5%, while the top performing decile is over five percentage points higher, at 37.6%. Figure 2 shows a comparison of the characteristics of the top decile to the rest of the fleet (deciles 1-9). While on average the top decile consisted of units with larger capacities, higher steam pressures, higher load factors, and a higher percentage burning bituminous coal, there was significant overlap with the rest of the fleet. For example, the average steam pressure at the turbine of the top decile is around 3000 psig, but there are ten units within the top decile with steam pressures in the 1800 psig to 2000 psig range. Also, the net nameplate capacity of the units in top decile, while larger on average, ranged from 114 MW to 1426 MW, indicating that even small plants can achieve higher than average efficiencies.

Figure 2 shows that the average age of the units in the top decile is roughly the same as the rest of the fleet, 40 years vs. 42 years, and both groupings contain units that are over 50 years old. Since coal generating units with SO_2 controls are expected to have lower efficiencies due to auxiliary load requirements, it is noteworthy that both the top decile and the rest of the fleet have almost the same percentage of units with sulfur controls.

Setting a CFPP Fleet Efficiency Target

Figure 3 shows 860 [2] coal generating units segmented by steam pressure [3], [4], coal type, and nameplate capacity [4]. If each segment is able to refurbish underperforming units to achieve a generation-weighted efficiency equal to that of the 90th percentile unit, then the generation-weighted CFPP fleet efficiency can be improved from 32.5% to 35.2% [5]. This efficiency target takes into account the differences between the top deci le and the rest of the fleet.

Decile	Number of Units	Capacity (GW)	Capacity Factor	2008 Total Generation (BkWh)	2008 Generation - Weighted Efficiency (HHV)
1	194	30.5	62%	165	27.6%
2	102	30.3	67%	179	29.9%
3	88	30.7	65%	176	30.8%
4	86	30.6	69%	185	31.6%
5	75	30.7	70%	189	32.2%
6	83	30.8	66%	178	32.9%
7	71	31.0	68%	186	33.8%
8	79	30.6	68%	183	34.7%
9	61	30.8	67%	181	35.7%
10	53	30.7	74%	201	37.6%
Overall	892	307	69%	1823	32.5%*

* Using data from EIA form 923 [3] for 2008 produces a generation weighted efficiency of 33.0%. The EV dataset is based on the EPA Continuous Emissions Monitoring System and results in lower efficiencies on average for each plant when compared to the EIA.

Figure 1. 2008 Generation Weighted CFPP Efficiency by Decile.

	2008 Top Decile		2008 Rest of Fleet	
	Average	Range	Average	Range
Generation Weighted Efficiency	37.6%	36.3% - 43.7%	32.0%	19.5% - 36.2%
Capacity (MW)	580	114 - 1426	329	27 - 1300
% of units that are Super Critical	55%	-	10%	-
Steam Pressure (psig)	2935	1800 - 3500	2088	600 - 5000
Age (yrs)	40	1 - 54	42	1 - 69
Load Factor*	83%	67% - 99%	75%	24% - 105%
% of units burning Bituminous Coal	66%	-	56%	-
% of units with SO_2 Controls	36%	-	35%	-

* Load factor = annual net generation / (net nameplate capacity * operating hours during the year). Operating hours is defined as the number of hours the unit is online and does not include the time the unit is shut down for maintenance or other reasons. Load factors may be artificially high for the top decile relative to the rest of the fleet because nameplate capacity corresponds to design efficiency not actual efficiency.

Figure 2. Characteristics of Top Decile and Rest of Fleet.

Segment Criteria			Sub-population Characteristics			Efficiency	
Unit Type	Coal Type	Size (MW)	Capacity (GW)	# Units	Generation (BkWh)	Average	90th 90 Percentile
Low Pressure Sub Critical (600 - 1600 psig)	Bit.		10.3	127	44	29.6%	33.0%
	Subbit.	0-200	4.6	59	26	27.5%	29.8%
	Other		0.6	7	2	27.4%	30.5%
High Pressure Sub Critical (1800 - 2600 psig)	Bit.	0-200	21.6	134	112	32.1%	34.8%
		200-500	33.4	103	189	32.8%	35.6%
		500+	29.7	48	176	32.7%	35.0%
	Subbit.	0-200	7.2	47	42	30.7%	32.5%
		200-500	31.2	97	191	31.4%	35.6%
		500+	64.4	98	401	31.6%	33.8%
	Other		11.1	28	72	31.7%	35.1%
Super Critical (3334+ psig)	Bit.		60.7	79	372	35.1%	37.3%
	Subbit.		15.0	20	90	35.2%	37.2%
	Other		8.1	13	55	31.8%	34.9%
Whole fleet			298	860	1772	32.5%	37.6%

Figure 3. Segmentation Analysis of the CFPP Fleet.

NETL is setting forth a vision for CFPP fleet-wide average efficiency of 36% based on (1) retirements of lower efficiency units combined with increased generation from higher efficiency refurbished units and (2) advanced refurbishments and improved operation and maintenance practices to increase average efficiency of some segments beyond the best-in-class.

The low pressure subcritical units and the 0-200 MW subbituminous units in Figure 3 had 90th percentile efficiencies that were significantly lower than the rest of the fleet. Retiring these units and relying on increased generation from the more efficient segments to maintain constant coal generation yields a fleet efficiency target of 35.6%.

Aggressive refurbishments could allow some segments to improve beyond the 90th percentile level. For example, the high pressure subcritical, subbituminous, 500 MW segment had a 90th percentile efficiency of 33.8% while the same segment in the 200-500 MW size had a 90th percentile of 35.6%.

Also, all of the high pressure subcritical and supercritical, bituminous coal segments achieved average efficiencies below the NETL bituminous baseline study [5] which sets an efficiency of 36.8% for a new subcritical CFPP and 39.1% for a new supercritical CFPP using existing technology.

Improving fleet efficiency beyond the 90th percentile level is also supported by analysis of historical efficiency data (1998-2008). NETL examined a scenario where each unit achieved its best efficiency over the past ten years in the same year. The units were ranked by best year efficiency and grouped into new deciles of equal capacity using the same method to produce Figure 1.

Figure 4 reports the generation-weighted average efficiency for the original 2008 decile analysis and the best year scenario. Note that deciles nine and ten (60 GW of capacity) achieve generation-weighted average efficiencies of 37.7% and 39.7% in their best years, which is higher than any individual 90th percentile segment. It is possible that units could routinely achieve the efficiency of their best year through better operation and maintenance practices [6].

Assuming that the CFPP fleet employs a combination of aggressive refurbishment and improved operation maintenance, 36% is a reasonable average efficiency target.

Decile	2008 Efficiency	1998 - 2008 Best Year Efficiency
1	27.6%	29.9%
2	29.9%	31.7%
3	30.8%	32.5%
4	31.6%	33.3%
5	32.2%	34.1%
6	32.9%	35.0%
7	33.8%	35.8%
8	34.7%	36.9%
9	35.7%	37.7%
10	37.6%	39.7%
Average	**32.5%**	**34.5%**

Figure 4. Best Year Generation-Weighted Efficiency Analysis.

New Source Review

The 36% CFPP fleet efficiency target does not take into account installing sulfur scrubbers at facilities that lack controls. New Source Review (NSR) requires that generating units over 73MW emit SO2 under 1.4lbs/MWh or reduce emissions to 90-95% of the possible emissions [6].The EV dataset showed that 250 GW of the 307 GW CFPP fleet emits SO2 above the 1.4lbs/MWh level, and that 165 GW lack any type of sulfur controls. If efficiency upgrades were done in conjunction with installing sulfur scrubbers on 165-250 GW of the fleet, the efficiency target would be reduced .5 to 1 percentage points.

Summary

The segmentation analysis serves as the basis for the 36% average CFPP fleet efficiency vision.

NETL expects that the fleet achieving efficiencies approaching the best-in-class will require a combination of improved operational and maintenance practices, capital investments in unit refurbishments, and application of advanced technologies.

Future Work

The analysis of the opportunity to increase CFPP efficiency could be improved by:

- Verification of coal generating unit efficiency data.
- Estimates of the cost of efficiency upgrades.
- Unit-specific data to enable estimation of the design heat rate for each generating unit. Case studies of efficiency upgrades at generating units, including ASPEN models, to provide concrete examples of the opportunity to improve efficiency.
- Detailed analyses of scenarios where some of the generating units are retired, some are refurbished and up-rated, and others are fixed with sulfur, NOx, and mercury controls. Analyses of how more efficient

coal plants will dispatch, what other generating sources will be displaced, and the overall effect on GHG emissions.

REFERENCES

[1] Department of Energy, Energy Information Administration, Annual Energy Outlook 2009 NEMS Run Results. http://www.eia. doe.gov/oiaf/servicerpt/stimulus/aeostim.html.

[2] NETL analysis of Ventyx's Velocity Suite 2008 net heat rate data for coal-fired units. Units with less than 10% capacity factor and using less than 97% coal were excluded from analysis. Units with missing or anomalous data were also excluded. The excluded units account for 3% of the electricity generated from coal in 2008. http://www.ventyx.com/velocity/vs-overview.asp

[3] Department of Energy, Energy Information Administration, Utility, Non-Utility, and Combined Heat & Power Plant Database. http://www.eia.doe.gov/cneaf/electricity/page/data.html

[4] 2008 Platts UDI World Electric Power Plant Database.

[5] Cost and Performance Baseline for Fossil Energy Plants Desk Reference (May 2007) http://www.netl.doe.gov/energy-analyses/pubs/ deskreference/index.html

[6] 40 CFR Part 60 – Standards of Performance for Electric Utility Steam Generating Units for Which Construction Commenced after September 18, 1978, Final Rule Amendments, February 27, 2006.

End Notes

[1] EV stores a collection of public data on power generating units that is primarily obtained from the Energy Information Administration (EIA) and the Environmental Protection Agency (EPA). This analysis uses the average annual net heat rate to calculate efficiency. Units with anomalous data, capacity factors under 10%, using less than 97% coal, and gasification plants, were removed from the dataset. The units that were removed from the dataset represented 3% of the total generation from all coal units. The resulting dataset contained 892 units.

[2] An additional 32 units were omitted from the segmentation analysis due to missing or anomalous steam pressure data.

[3] Steam pressure data was obtained from the 2008 Platts UDI database and combined with the dataset obtained from EV.

[4] In some cases units were not broken into capacity categories because the populations were small and mostly homogeneous. Although load factor for the top decile was an average of 83% compared to 75% for the rest of the fleet, it was deemed to be changeable through better dispatch.

[5] To calculate fleet efficiency target: 3412 / (Σ (3412/90th percentile eff. * generation) / 1772).

[6] NETL considered that units may achieve their best year efficiency when they are running at maximum load. To account for this we examined the correlation between best year efficiency and the year with the highest load factor and found no correlation. We also examined the years in which best year efficiency was achieved to see if there was any one year in which a preponderance of units achieved their best or worst year, and found that the years were evenly distributed.

In: Improving the Efficiency ...
Editor: Roland Ludovic
ISBN: 978-1-63117-590-9

Chapter 3

AVAILABLE AND EMERGING TECHNOLOGIES FOR REDUCING GREENHOUSE GAS EMISSIONS FROM COAL-FIRED ELECTRIC GENERATING UNITS*

United States Environmental Protection Agency

ACRONYMS AND ABBREVIATIONS

APFBC	Advanced pressurized fluidized bed combustion
ASTM	American Society for Testing and Materials
ASME	American Society of Mechanical Engineers
ASU	Air separation unit
BACT	Best Available Control Technology
Btu	British thermal unit
CAA	Clean Air Act
CCS	Carbon capture and storage
CEMS	Continuous emission monitoring system
CFB	Circulating fluidized bed
CH4	Methane

* This is an edited, reformatted and augmented version of the United States Environmental Protection Agency publication, dated October 2010.

CO	Carbon monoxide
CO_2	Carbon dioxide
EGU	Electric generating unit
EPA	U.S. Environmental Protection Agency
FBC	Fluidized bed combustion
EPRI	Electric Power Research Institute
FGD	Flue gas desulfurization
GHG	Greenhouse gas
H2O	Water
HRSG	Heat recovery steam generator
HHV	Higher heating value
IGCC	Integrated gasification combined cycle
IEA	International Energy Agency
kJ	Kilojoule
kW	Kilowatt
kWh	Kilowatt-hour
LCOE	Levelized cost of electricity
Mg	Megagram
MMBtu/hr	Million Btu per hour
MPa	Megapascal
MW	Megawatt
MWe	Megawatt electrical
MWh	Megawatt-hour
MSW	Municipal solid waste
N2O	Nitrous oxide
NETL	National Energy Technology Laboratory
NOX	Nitrogen oxides
O&M	Operation and maintenance
PC	Pulverized coal
PFBC	Pressurized fluidized bed combustion
PM	Particulate matter
PRB	Power River Basin
scfm	Standard cubic feet per minute
SO_2	Sulfur dioxide
SO_3	Sulfur trioxide
SNCR	Selective noncatalytic reduction
ton/day	tons per day
ton/yr	tons per year
U.S. DOE	U.S. Department of Energy

U.S. EIA U.S. Energy Information Administration

1. INTRODUCTION

This document is one of several white papers that summarize readily available information on control techniques and measures to mitigate greenhouse gas (GHG) emissions from specific industrial sectors. These white papers are solely intended to provide basic information on GHG control technologies and reduction measures in order to assist States and local air pollution control agencies, tribal authorities, and regulated entities in implementing technologies or measures to reduce GHGs under the Clean Air Act, particularly in permitting under the prevention of significant deterioration (PSD) program and the assessment of best available control technology (BACT). These white papers do not set policy, standards or otherwise establish any binding requirements; such requirements are contained in the applicable EPA regulations and approved state implementation plans.

This document provides information on control techniques and measures that are available to mitigate GHG emissions from the coal-fired electric generating sector at this time. The primary GHG emitted by the coal-fired electric generation industry is carbon dioxide (CO_2), and the control technologies and measures presented in this document focus on this pollutant. While a large number of available technologies are discussed here, this paper does not necessarily represent all potentially available technologies or measures that that may be considered for any given source for the purposes of reducing its GHG emissions. For example, controls that are applied to other industrial source categories with exhaust streams similar to the cement manufacturing sector may be available through “technology transfer” or new technologies may be developed for use in this sector.

The information presented in this document does not represent U.S. EPA endorsement of any particular control strategy. As such, it should not be construed as EPA approval of a particular control technology or measure, or of the emissions reductions that could be achieved by a particular unit or source under review.

1.1. Electric Power Generation Using Coal

Electricity is generated at most electric power plants by using mechanical energy to rotate the shaft of electromechanical generators. The mechanical energy needed to rotate the generator shaft can be produced from the conversion of chemical energy by burning fuels or from nuclear fission; from the conversion of kinetic energy from flowing water, wind, or tides; or from the conversion of thermal energy from geothermal wells or concentrated solar energy. Electricity also can be produced directly from sunlight using photovoltaic cells or by using a fuel cell to electrochemically convert chemical energy into an electric current.

In 2008, approximately 70% of the electricity used in the United States was generated by burning fossil fuels (coal, natural gas, petroleum liquids) (U.S. EIA 2010). The combustion of a fossil fuel to generate electricity can be either: 1) in a steam generating unit (also referred to simply as a "boiler") to feed a steam turbine that, in turn, spins an electric generator: or 2) in a combustion turbine or a reciprocating internal combustion engine that directly drives the generator. Some modern power plants use a "combined cycle" electric power generation process, in which a gaseous or liquid fuel is burned in a combustion turbine that both drives electrical generators and provides heat to produce steam in a heat recovery steam generator (HRSG). The steam produced by the HRSG is then fed to a steam turbine that drives a second electric generator. The combination of using the energy released by burning a fuel to drive both a combustion turbine generator set and a stream turbine generator significantly increases the overall efficiency of the electric power generation process.

Coal is the most abundant fossil fuel in the United States and is predominately used for electric power generation. In 2008, approximately 49% of the net electricity generated in the U.S. was produced by coal (U.S. EIA 2010). Historically, electric utilities have burned solid coal in steam generating units. However, coal can also be first gasified and then burned as a gaseous fuel. The integration of coal gasification technologies with the combined cycle electric generation process is called an integrated gasification combined cycle (IGCC) system or a "coal gasification facility". For the remainder of this document, the term "electric generating unit" or "EGU" is used to mean a solid fuel-fired steam generating unit that serves a generator that produces electricity for sale to the electric grid.

2. Coal-Fired Electric Generating Units

This section provides a summary overview of the types or ranks of coal that are typically burned in EGUs operating in the United States, the most commonly used combustion processes, and the resulting emissions of greenhouse gases.

2.1. Coals Burned in U.S. EGUs

In the United States, coals are ranked based on the degree of metamorphism (effectively, the geological age of the coal and the conditions under which the coal formed). These classification criteria have been standardized by the American Society for Testing and Materials (ASTM) method D-388. Under the ASTM method, coals are divided into four major categories called "ranks:" anthracite, bituminous coal, subbituminous coal, and lignite. Typical coal characteristics for the three most commonly used coal ranks are summarized in Exhibit 2-1.

Exhibit 2-1. Selected characteristics of major coal ranks used for electricity generation in the United States

Coal Ranka	Higher Heating Value (HHV) Range Defined by ASTM D-388	Typical Coal Moisture Content[b]	Coal Delivered for U.S. Electric Power Production in 2008[c,d]		
			Total Coal Quantity Delivered Nationwide (1,000 tons)	Average Ash Content	Average Sulfur Content
Bituminous	>10,500 Btu/lb	2 to 16%	463,943	10.6%	1.68%
Subbituminous	<10,500 Btu/lb and >8,300 Btu/lb	15 to 30%	522,228	5.8%	0.34%
Lignite	< 8,300 Btu/lb	25 to 40%	68,945	13.8%	0.86%

[a] Anthracite coal use is limited to reclaiming coal from coal refuse piles for use in a few power plants located close to the anthracite mines in eastern Pennsylvania.

[b] Reference: U.S. EPA, 2001.

[c] Reference: U.S. EIA, 2010, Table 3.6.

[d] Includes data collected from electric utilities, independent power producers, and combined heat and power producers.

Most coal-fired EGUs in the United States burn either bituminous or subbituminous coals. Approximately one half of the tonnage of coals delivered to U.S. electric power generation facilities was subbituminous (49.5%), and another 44% was bituminous coal. Some coal-fired EGUs burn multiple coal ranks. At many of these facilities, the coals are blended together before firing.

However, some facilities may switch between coal ranks because of site-specific considerations. The largest sources of bituminous coals burned in EGUs are mines in regions along the Appalachian Mountains, in southern Illinois, and in Indiana. Additional bituminous coals are supplied from mines in Utah and Colorado. The vast majority of subbituminous coals are supplied from mines in Wyoming and Montana, and many EGUs burn subbituminous coals from the Powder River Basin (PRB) region in Wyoming. This material is often referred to simply as "PRB coal."

In general, the burning of lignite or anthracite by electric utilities is limited to those EGUs that are located near the mines supplying the coal. Lignite accounted for approximately 6.5% of the total tonnage of coal delivered to electric utility power plants in 2008. All of those facilities were located near the coal deposits from which the lignite was mined in Texas, Louisiana, Mississippi, Montana, or North Dakota. Similarly, anthracite use was limited to a few power plants located close to the anthracite mines in eastern Pennsylvania. The coal-fired EGUs at those facilities primarily burn anthracite that has been reclaimed from coal refuse piles of previous mining operations. In general, "coal refuse" means any by-product of coal mining or coal cleaning operations with an ash content greater than 50 % (by weight) and a heating value less than 13,900 kilojoules per kilogram (kJ/kg) (6,000 Btu per pound (Btu/lb) on a dry basis. Coal refuse piles from previous mining operations are primarily located in Pennsylvania and West Virginia. Current mining operations generate less coal refuse than older ones.

2.2. Coal Utilization in U.S. EGUs

Steam turbine power plants operate on the Rankine thermodynamic cycle. The steam is produced by the boiler, where water pumped into the boiler ("feedwater") passes through a series of tubes to capture heat released by coal combustion and then boils under high pressure to become superheated steam. The superheated steam leaving the boiler then enters the steam turbine throttle, where it powers the turbine and connected generator to make electricity.

After the steam expands through the turbine, it exits the back end of the turbine into the surface condenser, where it is cooled and condensed back to water. This condensate is then returned to the boiler through high-pressure feed pumps for reuse. Heat from the condensing steam is normally rejected to cooling water circulated through the condenser which then goes to a surface water body, such as a river, or to an on-site cooling tower.

An EGU can be classified as either dry or wet bottom, depending on the ash removal technique used. Dry bottom boilers fire coals with high ash fusion temperatures, allowing for solid ash removal. In the less common wet bottom (slag tap) boilers, coal with a low ash fusion temperature is fired, and molten ash is drained from the bottom of the boiler.

Coal-fired EGUs use one of five basic coal utilization processes.

- Stoker-fired
- Pulverized coal (PC)
- Cyclone-fired
- Fluidized-bed combustion (FBC)
- Coal gasification (IGCC)

To improve the overall thermal conversion efficiency of the Rankine cycle, the majority of EGUs include a series of heat recovery sections. These sections are located downstream from the furnace chamber and are used to extract additional heat from the flue gas. The first section contains a "superheater," which is used to increase the steam temperature. The second heat recovery section contains a "reheater," which reheats the steam exhausted from the first stage of the steam turbine. This steam is then returned for another pass thorough a second stage of the turbine. The reheater is followed by an "economizer," which preheats the condensed feedwater recycled back to the boiler tubes in the furnace. The final heat recovery section is the "air heater," which preheats the ambient air used for coal combustion. The flue gas exhausted from the boiler passes through particulate matter (PM) and other air emissions control equipment before being vented to the atmosphere through a stack.

2.2.1. Stoker-Fired Coal Combustion

First introduced to the electric utility industry in the late 1800s, stoker-fired coal combustion is the oldest boiler coal-firing design. In a stoker-fired boiler, the coal is crushed and burned on a grate. Heated air passes upward through openings in the grate. Stokers are classified according to the way coal is fed to the grate – as underfeed stokers, overfeed stokers, and spreader stokers (see Exhibit 2-2). Stoker firing coal combustion is an obsolete technology for new coal-fired EGUs because the other newer coal combustion technologies provide superior coal combustion efficiency, applicability, and other advantages. There are still a few small stoker-fired EGUs in service in the U.S., but as these units are retired no new coal-fired stoker-fired EGUs are expected to be built. The majority of new stoker-fired boiler capacity is expected to occur at municipal solid waste combustor facilities and facilities burning solid biomass.

2.2.2. Pulverized-Coal Combustion

Pulverizing coal into a very fine powder allows the coal to be burned more easily and efficiently. For a PC-fired EGU, the coal must first be pulverized in a mill to the consistency of talcum powder (i.e., at least 70% of the particles will pass through a 200-mesh sieve). The pulverized coal is generally entrained in primary combustion air before being blown through the burners into the combustion chamber where it is fired in suspension. PC-fired boilers are classified by the firing position of the burners either as wall-fired or tangential-fired (see Exhibit 2-2).

A PC-fired boiler consists of multiple sections, and Exhibit 2-3 presents a simplified schematic of the major components of a PC-fired boiler using subcritical steam conditions. The pulverized coal is ignited and burned in the section of the boiler called the "furnace chamber" (or sometimes the "firebox"). Ambient air blown into the furnace chamber provides the oxygen required for combustion. The walls of the furnace chamber are lined with vertical tubes containing the feedwater. Heat transfer from the hot combustion gases in the furnace boils the water in the tubes to produce the high-temperature, high-pressure steam. The steam is separated from boiler water in a steam drum and sent to the steam turbine. The remaining water in the drum re-enters the boiler for further conversion to steam. The hot combustion products are vented from the furnace in a gas stream called collectively flue gas.

2.2.3. Cyclone Coal Combustion

Cyclone coal combustion technology was developed as an alternative to PC-firing because it requires less pre-processing of the coal and allows for the burning of lower rank coals with higher moisture and ash contents. Cyclone boilers use burner design and placement (i.e., several water-cooled horizontal burners) to produce high-temperature flames that circulate in a cyclonic pattern.

Exhibit 2-2. Characteristics of coal-firing configurations used for U.S. EGUs

Coal-firing Configuration	Application to U.S. EGUs	Coal Combustion Process Description	Distinctive Design/Operating Characteristics	
Stoker-fired	• Oldest coal-firing design first introduced to the electric utility industry in the late 1800s. • Not a significant contributor to overall U.S. nationwide MW generating capacity. • New EGUs are not expected to use this coal-firing design because of the superior performance and advantages of newer coal combustion technologies.	Coal is crushed into large lumps and burned in a fuel bed on a moving, vibrating, or stationary grate. Coal is pushed, dropped, or thrown onto the grate by a mechanical device called a "stoker."	Spreader-stoker	A flipping mechanism throws the coal into the furnace above the grate. The fine coal particles burn in suspension while heavier coal lumps fall to the grate and burn in a fuel bed.
			Underfeed	Coal fed by pushing the coal up underneath the burning fuel bed.
			Traveling grate	Coal is fed by gravity onto a moving grate and leveled by a stationary bar at the furnace entrance.
Pulverized-Coal Combustion	• Coal-firing design predominately used at existing U.S. EGUs • In 2008, consumed ~ 92% of total coal consumed by U.S. EGUs.[a] • Currently coal-firing design of choice for new large coal-fired EGUs (> 400 MWe) built in U.S.	Coal is ground to a fine powder that is pneumatically fed to a burner where it is mixed with combustion air and then blown into the furnace. The pulverized- coal particles burn in suspension in the furnace. Unburned and partially burned coal particles are carried off with the flue gas.	Wall-fired	An array of burners fire into the furnace horizontally, and can be positioned on one wall or opposing walls depending on the furnace design.
			Tangential-fired (Corner-fired)	Multiple burners are positioned in opposite corners of the furnace producing a fireball that moves in a cyclonic motion and expands to fill the furnace.

Exhibit 2-2. (Continued)

Coal-firing Configuration	Application to U.S. EGUs	Coal Combustion Process Description	Distinctive Design/Operating Characteristics	
Cyclone	• Existing cyclone EGUs in U.S. constructed prior to 1981. • In 2008, consumed ~ 6% of total coal consumed by U.S. EGUs. • New EGUs are not expected to use this boiler type because of the commercial availability of FBC technology.	Coal is crushed into small pieces and fed through a burner into the cyclone furnace. A portion of the combustion air enters the burner tangentially creating a whirling motion to the incoming coal.	Designed to burn coals with low-ash fusion temperatures that are difficult to burn in PC boilers. The majority of the ash is retained in the form of a molten slag.	
Fluidized-bed Combustion	• FBC EGUs increasingly being built in U.S. to burn low rank coals, coal refuse, and blends of coal with other solid fuels such as petroleum coke or biomass. • In 2008, consumed approximately 2% of total coal consumed by U.S. EGUs.a • Atmospheric FBC EGUs are currently operating in the U.S. with generating capacities in the range of 250 to 300 MWe. • No Pressurized FBC boilers currently used for U.S. EGUs	Coal is crushed into fine particles. The coal particles are suspended in a fluidized bed by upward-blowing jets of air. The result is a turbulent mixing of combustion air with the coal particles. Typically, the coal is mixed with a sorbent such as limestone (for SO_2 emission control). The unit can be designed for combustion within the bed to occur at atmospheric or elevated pressures. Operating temperatures for FBC are in the range of 1,500 to 1,650°F (800 to 900°C).	Bubbling fluidized bed (BFB)	Operates at relatively low gas stream velocities and with coarse-bed size particles. Air in excess of that required to fluidize the bed passes through the bed in the form of bubbles.
			Circulating fluidized bed (CFB)	Operates at higher gas stream velocities and with finer-bed size particles. No defined bed surface. Must use high-volume, hot cyclone separators to recirculate entrained solid particles in flue gas to maintain the bed and achieve high combustion efficiency.

Coal-firing Configuration	Application to U.S. EGUs	Coal Combustion Process Description	Distinctive Design/Operating Characteristics
Coal Gasification (e.g., IGCC)	• Limited application to EGUs to date. • Some new proposed EGU projects using coal gasification as part of IGCC plant.	Synthetic combustible gas ("syngas") derived from an on-site coal gasification process is burned in a combustion turbine. The hot exhaust gases from the combustion turbine pass through a heat recovery steam generator to produce steam for driving a steam turbine/generator unit.	Coal gasification units are unique from the other coal-firing configurations because a gaseous fuel (synfuel or syngas) is burned instead of solid coal and combines the Rankine and Brayton thermodynamic cycles as is the case for a combined cycle power plant.

[a] Source: U.S. EIA, 2008.

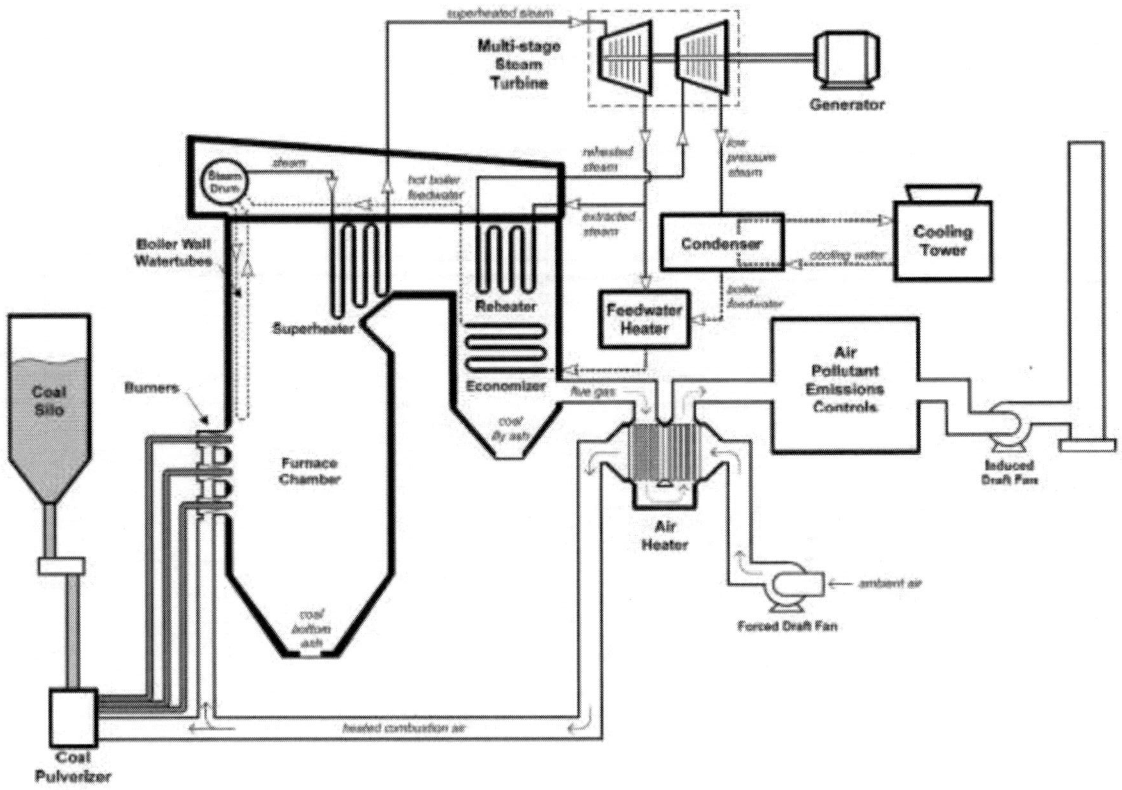

Exhibit 2-3. Simplified schematic of a PC-fired EGU using a subcritical boiler.

The coal is crushed to a 4-mesh size, and then fed tangentially with primary air, to a horizontal cylindrical combustion chamber. In this chamber, small coal particles are burned in suspension, while the larger particles are forced against the outer wall. The high temperatures developed in the relatively small boiler volume, combined with the low fusion temperature of the coal ash, causes the ash to form a molten slag, which is drained from the bottom of the boiler through a slag tap opening. Existing cyclone EGUs in the U.S. were designed or installed before 1981. Cyclone EGUs have high nitrogen oxides (NOx) emission rates and no new cyclone boilers are expected to be built. Fluidized-bed combustion is an alternative technology that is able to burn lower rank coals without high NOx emissions.

2.2.4. Fluidized-Bed Combustion

The term "fluidized" refers to the state of the bed materials (fuel and inert material [or sorbent]) as gas passes through the bed. In a typical FBC EGU, combustion occurs when coal and a sorbent, such as limestone, are suspended through the action of primary combustion air distributed below the combustor floor. The gas cushion between the solids allows the particles to move freely, giving the bed a liquid-like characteristic (i.e., fluidized). FBC can occur in either atmospheric or pressurized boilers. Two fluidized bed designs can be used for atmospheric and pressurized FBC boilers: a bubbling fluidized bed or a circulating fluidized bed (CFB) (see Exhibit 2-2). An advantage of CFB boiler EGUs compared to PC-fired EGUs is fuel flexibility. A CFB boiler EGU can burn any rank of coal (including coal refuse), petroleum coke (a carbonaceous solid derived from oil refinery coker units or other cracking processes), and biomass without significant modifications.

The combustion temperature of a FBC boiler (1,500 to 1,650°F) is significantly lower than a PC-fired boiler (2,450 to 2,750°F), which results in lower NOX formation and the ability to capture sulfur dioxide (SO_2) with limestone injection in the furnace. Even though the combustion temperature of a FBC boiler is low, the circulation of hot particles provides efficient heat transfer to the furnace walls and allows longer residence time for carbon combustion and limestone reaction. This results in good combustion efficiencies, comparable to PC-fired EGUs.

Atmospheric CFB boilers have successfully been scaled-up and are operating at a number of facilities throughout the world. Exhibit 2-4 presents a simplified schematic of the major components of a CFB boiler EGU. Calcium in the sorbent combines with SO_2 gas to form calcium sulfite and sulfate solids, and solids exit the combustion chamber and flow into a hot cyclone.

The cyclone separates the solids from the gases, and the solids are recycled for combustor temperature control. Heat in the flue gas exiting the hot cyclone is recovered in a series of heat recovery sections of the boiler to produce steam. The superheated steam leaving the boiler then enters the steam turbine, which powers a generator to produce electricity. Like PC-fired EGUs, CFB boilers can be used with either subcritical or supercritical steam cycles.

Currently, the capacity of CFB subcritical boilers ranges from 25 to 350 MWe. Examples of these systems include (Foster Wheeler North America Corp., 2009):

- Two 300 MWe CFB subcritical boilers at the Jacksonville Energy Authority power plant in Jacksonville, Florida. These units are capable of burning either 100% coal or 100% petroleum coke or any combination of the two.
- Three 262 MWe CFB subcritical boilers at the Turow power plant in Poland. The fuel for these boilers is lignite with moisture content of 45% by weight.

The largest atmospheric CFB boiler in operation to date is a 460 MWe unit at a power plant owned by the Polish utility company Południowy Koncern Energetyczny SA (PKE) in Lagisza, Poland (Foster Wheeler North America Corp., 2009). This unit is also the world's first supercritical CFB boiler. The primary fuel burned in the unit is Polish bituminous coal. The commercial operation of this unit demonstrates the successful integration of CFB boiler technology with supercritical boiler technology. The unit features include a vertical evaporator with supercritical steam conditions (4,000 psia, 1,050/1,075°F) and a reported overall net plant efficiency of 41.6% (HHV basis). Based on the design and operating experience with the Lagisza Power Plant, both 600 and 800 MWe size supercritical CFB boiler designs with full commercial guarantees are being offered (Foster Wheeler North America Corp., 2009).

Pressurized fluidized-bed combustion (PFBC) systems are FBC systems that operate at elevated pressures (typically pressures of 1-1.5 MPa) and produce a high-pressure gas stream at temperatures that can drive a turbine. As with atmospheric FBC, two formats are possible, one with bubbling beds, the other with a circulating configuration. Currently, all operating units use bubbling beds. In a PFBC, the combustor and hot gas cyclones are all enclosed in a pressure vessel. Both coal and sorbent (for SO_2 emissions reductions) have to be fed across the pressure boundary, and similar provision for ash removal is necessary. For hard coal (i.e., bituminous coal) applications, the coal and limestone can be crushed together, and then fed as a paste, with 25% water. As with atmospheric FBC, a combustion temperature between 1,500 to 1,650°F (800 to 900°C) has the advantage of less NOX formation than in PC combustion. In addition, the effectiveness of a CCS system is increased due to the high pressure within the PFBC cycle and higher partial pressure of the CO_2 in the hot gas stream.

The initial or first-generation PFBC designs are based on directly burning crushed coal in the combustor. The high pressure gas is first expanded through a turbine and then heat is recovered from the turbine exhaust in a HRSG to produce steam, which is used to drive a conventional steam turbine. Exhibit 2-5 presents a simplified schematic of the major components of a PFBC EGU. A number of demonstration projects (ranging in size from 60 to 130 MWe) were conducted during the 1990s in Japan, Spain, Sweden, the United Kingdom, the U.S., and other countries. Japanese equipment manufacturers and electric power companies have led the commercial development of PFBC technology with the construction of several commercial-scale units.

- 360 MWe PFBC unit operated by Kyushu Electric Power Company at the Karita Power Station located near Kitakyushu, Japan. The unit began commercial operation in July 2001. The unit uses a supercritical boiler and has a reported net efficiency based on test results of 41.8% HHV (Asai, 2004).
- 250 MWe PFBC unit operated by Chugoku Electric Power Co., Inc. at Osaki Power Station located near Hiroshima, Japan. Unit 1 began commercial operation in 2000. The planned construction of a second PFBC unit at the facility was cancelled in 2008

More advanced second-generation PFBC system designs use a pressurized carbonizer to first process the feed coal into fuel gas and char (solid material that remains after light gases and tar have been driven-out during the initial stage of combustion). The PFBC burns the char to produce steam and to heat combustion air for the combustion turbine. The fuel gas from the carbonizer burns in a topping combustor linked to a combustion turbine, heating the gases to the rated firing temperature of the combustion turbine. Heat is recovered from the combustion turbine exhaust in a HRSG to produce steam, which is used to drive a conventional steam turbine. These systems are also called advanced circulating pressurized fluidized-bed combustion (APFBC) combined cycle systems.

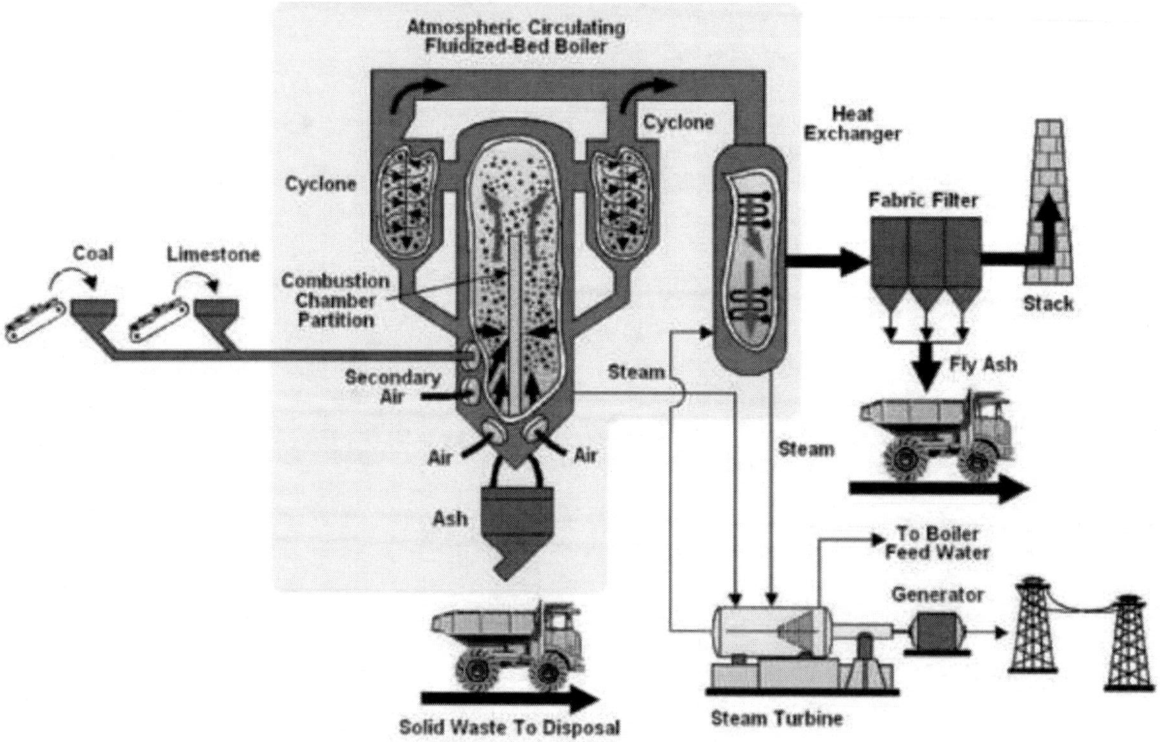

Source: NETL, 2010b, CCPI/Clean Coal Demonstrations Nucla CFB Demonstration Project, Project Fact Sheet.

Exhibit 2-4. Simplified schematic of an atmospheric circulating fluidized-bed (CFB) boiler power plant.

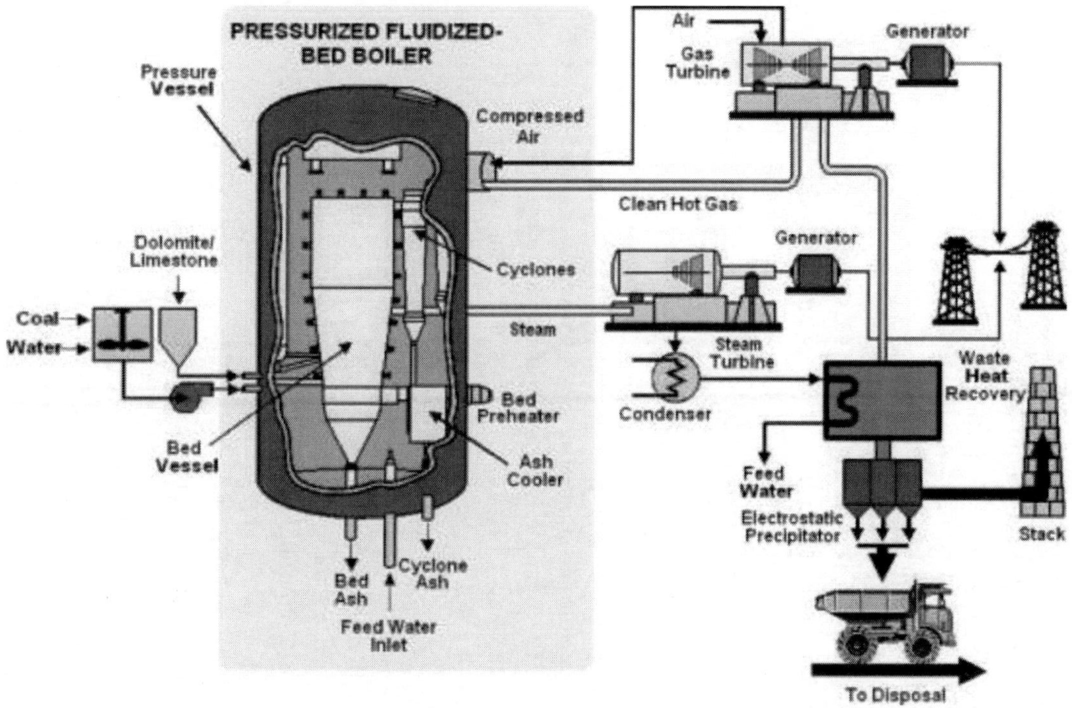

Source: NETL, 2010c, CCPI/Clean Coal Demonstrations Tidd PFBC Demonstration Project, Project Fact Sheet.

Exhibit 2-5. Simplified schematic of a pressurized fluidized-bed combustion (PFBC) power plant.

2.2.5. Coal Gasification

An IGCC power plant uses a coal gasification system to convert coal into a synthetic gas, which is then used as fuel in a combined cycle electric generation process. Coal is gasified by a process in which coal or a coal/water slurry is reacted at high temperature and pressure with oxygen (or air) and steam in a vessel referred to as a “gasifier” to produce a combustible gas composed of a mixture of carbon monoxide (CO) and hydrogen. This gas is often referred to as synthetic gas or syngas. Gasification processes have been developed using a variety of designs including moving bed, fluidized bed, entrained flow, and transport gasifiers. Coal gasification processes are offered by a number of companies with varying degrees of existing commercial application (NETL, 2010a). Exhibit 2-6 presents a simplified schematic of the major components of an IGCC power plant. The hot syngas can then be processed to remove sulfur compounds, mercury, and PM before it is used to fuel a combustion turbine generator to produce electricity. The heat in the exhaust gases from the combustion turbine is recovered to generate additional steam. This steam, along with the steam produced by the gasification process, then drives a steam turbine generator to produce additional electricity.

The efficiency of an IGCC power plant is comparable to the latest advanced PC-fired and CFB EGU designs using supercritical boilers. The advantages of using IGCC technology can include greater fuel flexibility (e.g., capability to use a wider variety of coal ranks), potential improved control of PM, SO_2 emissions, and other air pollutants, with the need for fewer post-combustion control devices (e.g., almost all of the sulfur and ash in the coal can be removed once the fuel is gasified and prior to combustion), generation of less solid waste requiring disposal, and reduced water consumption when compared to an EGU using a supercritical boiler (U.S. EPA, 2006). Disadvantages of using IGCC include additional plant complexity, higher construction costs, and poorer performance at high altitude locations when compared to an EGU using a supercritical boiler. However, IGCC power plants offer the potential for lower control costs of CO_2 emissions because the CO_2 in the syngas can be removed prior to combustion. Interest by U.S. electric utilities in building new IGCC power plants is increasing because of site-specific considerations and potential cost benefits for the technology.

Currently operating IGCC plants include the following U.S. and foreign plants (NETL, 2010a):

- 253 MWe IGCC plant at the NUON Willem-Alexander Power Plant in Buggenum, Netherlands. The unit began operation in 1994.

- 262 MWe IGCC plant at the Duke Energy Wabash River Power Station in Indiana. The unit began operation in 1995.
- 250 MWe IGCC plant at the Tampa Electric Company (TECO) Polk Power Station in Florida. The unit began operation in 1996.
- 400 MWe IGCC plant at the SUV power plant in Vresova, Czech Republic. The unit began operation in 1996.
- 283 MWe IGCC plant at the ELCOGAS power plant in Puertollano, Spain. The unit began operation in 1998.
- 250 MWe IGCC plant at the Joban Joint Electric Power Company Nakoso Power Station in Iwaki City, Japan. The unit began operation in 2007.

Over the past 5 years, a number of larger IGCC power plant projects have been proposed by U.S. electric utility companies. Some of these IGCC projects have been indefinitely delayed or canceled because of economic and regulatory factors, such as escalating project investment costs beyond initial estimates and unresolved cost recovery issues with State public utility commissions. One commercial IGCC project currently under construction is a 630 MWe IGCC facility at the Duke Energy Edwardsport Power Station in Knox County, Indiana.

Syngas produced by coal gasification can not only be used as a fuel to generate electricity or steam but also as a basic chemical building block for a large number of petrochemical and refining products. Because of these multiple uses, future IGCC projects may include facilities that integrate electricity generation with the production of other industrial outputs such as chemical feedstocks for manufacturing operations or hydrogen fuel for vehicles.

2.3. GHG Emissions from Coal-Fired EGUs

The principal chemical constituents of coal are carbon, hydrogen, oxygen, nitrogen, sulfur, moisture, and incombustible mineral matter (i.e., ash). When coal is burned, the carbon and hydrogen are oxidized to form the primary combustion products of CO_2 and water. Other combustion products such as NOX, SO_2, CO, and PM are formed in varying amounts.

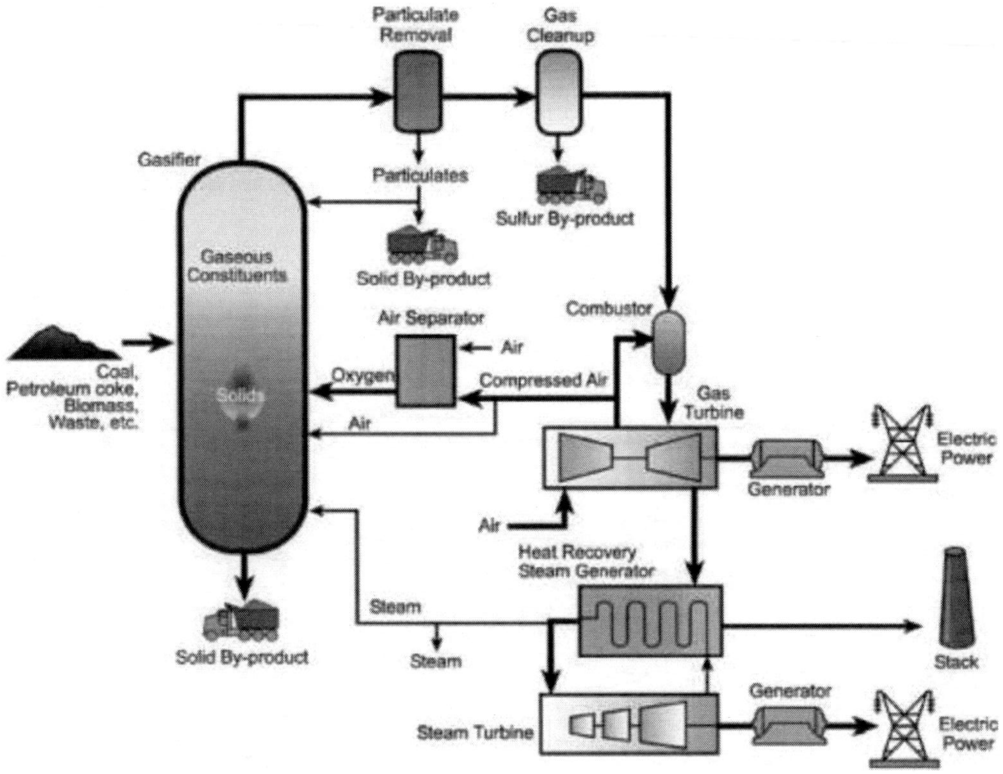

Source: NETL, 2010a, Overview of DOE's Gasification Program.

Exhibit 2-6. Simplified schematic of an integrated gasification combined cycle (IGCC) power plant.

The principal GHGs that enter the atmosphere because of human activities are CO_2, nitrous oxide (N_2O), methane (CH4), hydrofluorocarbons (HFC's), perfluorocarbons (PCF's), and sulfur hexafluoride (SF6). Of these, CO_2 is by far the most abundant GHG emitted from power production by coal utilization. To optimize overall efficiency for a given EGU, the unit is operated under conditions such that nearly all of the fuel carbon is converted to CO_2 during the combustion process. Methane is emitted during the mining and transport of coal but is not a significant by-product of EGU coal combustion. Fluorinated gases are not formed by coal combustion. Sulfur hexafluoride might be used at the power plant switchyard, but the switchyard is not typically considered part of the EGU.

Formation of N_2O during the combustion process results from a complex series of reactions and its formation is dependent upon many factors. However, the formation of N_2O is minimized when combustion temperatures are kept high and excess air is kept to a minimum. PC-fired EGUs are typically operated at these conditions and are not significant sources of N2O emissions. However, FBC EGUs can have measurable N2O emissions, resulting from the lower combustion temperatures and the use of selective noncatalytic reduction (SNCR) to reduce NOX emissions. Operating factors impacting N2O formation include combustion temperature, excess air, and sorbent feed rate (Korhonen, 2001). The N2O formation resulting from SNCR depends upon the reagent used, the amount of reagent injected, and the injection temperature (Weijuan, 2007).

2.4. Factors Impacting Coal-Fired EGU CO_2 Emissions

The level of CO_2 emissions that can potentially be released from a given coal-fired EGU depends on the type of coal burned, the overall efficiency of the power generation process, and use of air pollution control devices.

2.4.1. Impact of Coal Rank on CO_2 Emissions from EGUs

The amount of CO_2 that potentially can be emitted from a coal-fired EGU varies depending on the coal rank burned. The amount of heat released by coal combustion depends on the amounts of carbon, hydrogen, and oxygen present in the coal and, to a lesser extent, on the sulfur content. Hence, the ratio of carbon to heat content depends on these heat-producing components of coal, and these components vary by coal rank. **Exhibit 2-7** presents a comparison of the CO_2 emissions for the average heating values of U.S. coals. The values presented in the table are arithmetic averages and assume complete combustion. Based on these averages, in general anthracite emits the largest amount of CO_2 per million Btu (MMBtu), followed by lignite, subbituminous coal, and bituminous coal. However, for a given coal rank there is variation in the CO_2 emission factor depending on the coal bed from which the coal is mined.

Exhibit 2-7. CO_2 emission factors for coal by coal rank

Coal Rank	CO_2 Emissions per Unit of Heat Input (lbs CO_2/MMBtu)	
	U.S. Average	Range Across States with Coal Rank Deposits
Anthracite	227.4	227.4
Bituminous	205.3	201.3 to 211.6
Subbituminous	211.9	207.1 to 214.0
Lignite	216.3	211.7 to 220.6

Source: U.S. EIA (Hong, R. and E. Slatick, 1994).

In addition to the lower CO_2 emissions rate per unit of heat input (lbs CO_2/MMBtu), due to the inherent moisture in subbituminous and lignite coals, all else being equal a bituminous coal-fired boiler is more efficient than a corresponding boiler burning subbituminous or lignite coal. Therefore, switching from a low to a high-rank coal will tend to lower GHG emissions from the utility stack. However, overall GHG emissions might not be lowered by switching to bituminous coal. All coal mining operations release coal bed methane to the atmosphere during the mining process. Some bituminous coal reserves release significant amounts of methane, which could, in theory, offset GHG savings. Additional factors when considering overall GHG emissions include the fuel needs to mine, process, and transport the coal.

Additional solid fuels burned in EGUs include petroleum coke, biomass, and municipal solid waste (MSW). Petroleum coke has one of the highest CO_2 emissions rate (225 lb CO_2/MMBtu) of commonly used solid fuels. MSW combustors provide significant GHG reductions as an alternate to landfills (Kaplan, 2008). However, due to the difficulties associated with transporting large amounts of solid waste, MSW combustor facilities used for electrical power generation are typically limited to less than 100 MW of electrical output.

Of the gaseous and liquid fossil fuels used in steam generating units, natural gas combustion releases approximately 117 pounds of CO_2 per MMBtu, distillate oil releases 161 lb CO_2/MMBtu, and residual oil releases 174 lb CO_2/MMBtu. However, none of these fuels are typically used in new baseload steam generating units (e.g., boilers). Natural gas and distillate oil are significantly more expensive per unit heat input than coal. In addition, combustion turbines burning natural gas and distillate oil generate power more efficiently than a boiler burning natural gas and distillate oil. New baseload electric generation based on the use of either natural gas or distillate would likely use combined cycle combustion turbines. Therefore, aside from small amounts of natural gas for startup, shutdown, and potentially for combustion control, few new steam generating units are expected to burn significant quantities of either of these fuels directly in the boiler. Existing EGUs that burn natural gas and distillate oil tend to be older units that operate in a peaking or cycling mode. However, several base load coal-fired EGUs have been converted to natural gas. Natural gas-fired boilers tend to be less efficient than coal-fired boilers; however, they can startup and change loads more quickly than similar coal-fired boilers, do not typically require post combustion controls, and the fuel handling is simpler. Residual oil also tends to be more expensive than coal per unit of heat input, and because post-combustion environmental controls would still often be required, it is also not a common fuel choice for EGUs in the Lower 48 States. There has not been a new residual oil-fired EGU built in the Lower 48 States since 1981.

2.4.2 Impact of Coal-Fired EGU Efficiency on CO_2 Emissions

As the thermal efficiency of a coal-fired EGU is increased, less coal is burned per kilowatt-hour (kWh) generated, and there is a corresponding decrease in CO_2 and other air emissions. There is no standardized procedure for continuous on-line measurement of coal-fired EGU thermal efficiency (Peltier, 2010). However, a near approximation performed under EPA's Acid Rain Program collects heat input and gross megawatt output on an hourly basis to calculate gross heat rate. The heat input is derived from standardized continuous emission monitors, while the utility supplies gross megawatt output. The electric energy output as a fraction of the fuel energy input expressed in percentage is a commonly-used practice for reporting the efficiency of a coal-fired EGU. The greater the output of electric energy for a given amount of fuel energy input, the higher the efficiency for the electric generation process. Heat rate is another common way to express efficiency. Heat rate is expressed as the number of Btu or kJ required to generate a kWh of electricity. Lower heat rates are associated with more efficient power generating plants. Although the same basic formula is used to calculate efficiency for coal-fired EGUs, there are different methodologies for measuring the appropriate parameters. For example, the varying accuracy of the different methodologies can cause discrepancies in the measurement the heating value of the coal burned.

Efficiency can be calculated using the higher heating value (HHV) or the lower heating value (LHV) determined for the fuel. The HHV is the heating value directly determined by calorimetric measurement of the fuel in the laboratory. The LHV is calculated using a formula to account for the moisture in the fuel (i.e., subtract the energy required to vaporize the water in the coal and is thus not available to produce steam) and is a smaller value than the HHV. Consequently, the HHV efficiency for a given EGU is always lower than the corresponding LHV efficiency, because the reported heat input is larger. For bituminous coals the HHV efficiency value is typically about 2 percentage points lower than the corresponding LHV efficiency. For higher moisture subbituminous coals and lignites, the HHV efficiency is approximately 3 to 5 percentage points lower than the corresponding LHV efficiency (depending on moisture content). In engineering practice, HHV is typically used in the U.S. to express the efficiency of steam electric power plants while in Europe the practice is to use LHV. Similarly, the electric energy output for an EGU can be expressed as either of two measured values. One value relates to the amount of total electric power generated by the EGU, or "gross output." However, a portion of this electricity must be used by the EGU facility to operate the unit, including pumps, fans, electric motors, and pollution control equipment. This in-facility electrical load, often referred to as the "parasitic load," reduces the amount of power that can be delivered to the transmission grid for distribution and sale to customers. Consequently, electric energy output is also expressed in terms of "net output," which reflects the EGU gross output minus its parasitic load. When using efficiency to compare the effectiveness of different coal-fired EGU configurations and the applicable GHG emissions control technologies, it is important to ensure that all efficiencies are calculated using the same type of heating value (i.e., HHV or LHV) and the same type of electric energy output (i.e., gross MWh or net MWh). Although there is a direct inverse correlation between coal-fired EGU efficiency and CO_2 emissions, other factors must be considered when comparing the effectiveness of GHG control technologies to improve the efficiency of a given coal-fired EGU. The actual overall efficiency that a given coal-fired EGU achieves is determined by the interaction of a combination of site-specific factors that impact efficiency to varying degrees. These factors include:

- *EGU thermodynamic cycle* – EGU efficiency can be significant improved by using a supercritical or ultra-supercritical steam cycle.
- *EGU coal rank and quality* – EGUs burning higher quality coals (e.g., bituminous) tend to be more efficient than EGUs burning lower quality coals (e.g., lignite).
- *EGU plant size* – The electric-generating capacity of EGUs ranges from approximately 25 to 1,300 MWe. Assuming an EGU efficiency of 33% (a typical efficiency for existing coal-fired EGUs), this corresponds to a heat input range of 250 to 13,400 MMBtu/hr. EGU efficiency generally increases with size because the boiler and steam turbine losses are lower for larger equipment. However, as equipment size increases the differences in these losses start to taper off.
- *EGU pollution control systems* – The electric power consumed by air pollution control equipment reduces the overall efficiency of the EGU.
- *EGU operating and maintenance practices* – The specific practices used by an individual electric utility company for combustion optimization, equipment maintenance, etc. can affect EGU efficiency.
- *EGU cooling system* – The temperature of the cooling water entering the condenser can have impacts on steam turbine performance. Once-through cooling systems can have an efficiency advantage over recirculating cooling systems (e.g., cooling towers). However, once-though cooling systems typically have larger water related ecological concerns than recirculating cooling systems.
- *EGU geographic location* – The elevation and seasonal ambient temperatures at the facility site potentially may have a measureable impact on EGU efficiency. At higher elevations, air pressure is lower and less oxygen is available for combustion per unit volume of ambient air than at lower elevations. Cooler ambient temperatures theoretically could increase the overall EGU efficiency by increasing the draft pressure of the boiler flue gases and the condenser vacuum, and by increasing the efficiency of a condenser recirculating cooling system.
- *EGU load generation flexibility requirements* – Operating an EGU as a baseload unit is more efficient than operating an EGU as a load cycling unit to respond to fluctuations in customer electricity demand.

- *EGU equipment manufacturers* – The efficiency specifications of major EGU components such as boilers, turbines, and electrical generators provided by equipment manufacturers can affect EGU efficiency.
- *EGU plant components* – EGUs using the optimum number of feedwater heaters, high-efficiency electric motors, variable speed drives, better materials for heat exchangers, etc. tend to be more efficient.

Because of these factors, coal-fired EGUs that are identical in design but operated by different utility companies in different locations may have different efficiencies. Thus, the level of effectiveness of a given GHG control technology used to improve the efficiency at one coal-fired EGU facility may not necessarily directly transfer to a coal-fired EGU facility at a different location.

2.4.3. Impact of SO_2 Controls on Coal-Fired EGU CO_2 Emissions

The SO_2 emissions from new coal-fired EGUs, or retrofitting of an existing facility without specific SO_2 controls, are controlled using flue gas desulfurization (FGD) technology to remove the SO_2 before it is vented to the atmosphere. The selection of the type of FGD technology will impact overall GHG emissions. All FGD processes require varying amounts of electric power to operate, which contributes to the overall parasitic load of the unit. The FGD parasitic load requirements are typically between 1-2% of the gross output of the facility. In addition, some FGD processes use carbon-containing reagents (e.g., carbonates) that form CO_2 as a byproduct of the chemical reactions of the reagent with SO_2.

For a typical unit, the CO_2 that is chemically created in a scrubber adds an additional 1% to the overall GHG emissions, but it can be as high as 3% for facilities burning high sulfur coals. However, from an overall GHG emissions standpoint the use of FGD technologies that do not form byproduct CO_2, such as lime-based scrubbers, do not necessarily reduce emissions. Lime is manufactured by heating limestone in the absence of oxygen to remove a molecule of CO_2 ($CaCO_3$ + heat $\rightarrow$ $CaO + CO_2$). Unless the CO_2 is sequestered at the lime production facility, overall GHG emissions will be similar. A list of FGD processes used for controlling SO_2 emissions from coal-fired EGUs is presented in Exhibit 2-8, identifying those processes that chemically form additional CO_2.

Exhibit 2-8. CO_2 formation from coal-fired EGU flue gas desulfurization (FGD) processes

FGD Type	Reagent	Forms CO_2	Overall Reaction(s)	Reference
Wet Scrubbing	Limestone ($CaCO_3$)	yes	$CaCO_3 + SO_2 + \frac{1}{2}H_2O \rightarrow CaSO_3 \bullet \frac{1}{2}H_2O(s) + CO_2$; $CaCO_3 + SO_2 + 2H_2O + \frac{1}{2}O_2 \rightarrow CaSO_4 \bullet 2H_2O(s) + CO_2$	Ref 1
	Magnesium-enhanced lime; Dolomitic lime [$Ca(OH)_2 \bullet Mg(OH)_2$]	no	$10Ca(OH)_2 + 11SO_2 + Mg(OH)_2 \rightarrow 10CaSO_3 \bullet \frac{1}{2}H_2O(s) + MgSO_3 + 6H_2O$	Ref 1,4
	Dual Alkali; Sodium solution and lime	no	$2NaOH + SO_2 \rightarrow Na_2SO_3 + H_20$; $H_2O + Na_2SO_3 + SO_2 \rightarrow 2NaHSO_3$; $2NaHSO_3 + Ca(OH)_2 \rightarrow Na_2SO_3 + CaSO_3 \bullet \frac{1}{2}H_2O + 3/2H_2O$; $Na_2SO_3 + Ca(OH)_2 \rightarrow 2NaOH + CaSO_3$	Ref 2
	Dual Alkali (Dowa)	yes	$Al_2O_3 \bullet Al_2(SO4)_3 + 3SO_2 + 3/2O_2 \rightarrow 2Al_2(SO4)_3$; $2Al_2(SO4)_3 + 3CaCO_3 \rightarrow Al_2O_3 \bullet Al_2(SO4)_3 + 2CaSO_4(s) + 3CO_2$	Ref 3
	Seawater	yes	$2NaHCO_3 + SO_2 \rightarrow Na_2SO_3 + 2CO_2 + H_20$; $Na_2SO_3 + \frac{1}{2}O_2 \rightarrow Na_2SO_4$	Ref 3
	Magnesium oxide (MgO)	no	$MgO + SO_2 \rightarrow MgSO_3$	Ref 7
	Hydrogen Peroxide (H_2O_2)	no	$H_2O_2 + SO_2 \rightarrow H_2SO_4$	Ref 3
	Sodium hydroxide (NaOH)	no	$2NaOH + SO_2 \rightarrow Na_2SO_3 + H_20$; $Na_2SO_3 + \frac{1}{2}O_2 \rightarrow Na_2SO_4$	Ref 3

Exhibit 2-8. (Continued)

FGD Type	Reagent	Forms CO_2	Overall Reaction(s)	Reference
Dry/Semi-dry Scrubbing	Hydrated calcitic lime ($Ca(OH)_2$)	no	$Ca(OH)_2 + SO_2 \rightarrow CaSO_3 \bullet \frac{1}{2}H_2O(s) + \frac{1}{2}H_2O$; $Ca(OH)_2 + SO_2 + H_2O + \frac{1}{2}O_2 \rightarrow CaSO_4 \bullet 2H_2O(s)$	Ref 1
	Sodium bicarbonate ($NaHCO_3$)	yes	$2NaHCO_3 + SO_2 \rightarrow Na_2SO_3 + 2CO_2 + H_20$; $2NaHCO_3 + SO_2 + \frac{1}{2}O_2 \rightarrow Na_2SO_4 + 2CO_2 + H_20$	Ref 8
	Sodium sesquicarbonate (trona)	yes	$2(Na_2CO_3 \bullet NaHCO_3 \bullet 2H_2O) + 3SO_2 \rightarrow 3Na_2SO_3 + 5H_2O + 4CO_2$; $2(Na_2CO_3 \bullet NaHCO_3 \bullet 2H_2O) + 3SO_2 + 3/2O_2 \rightarrow 3Na_2SO_4 + 5H_2O + 4CO_2$	Ref 8
	Sodium carbonate (soda ash, Na_2CO_3)	yes	$Na_2CO_3 + SO_2 + \frac{1}{2}O_2 \rightarrow Na_2SO_4 + CO_2$	Ref 8
	Pulverized limestone	yes	$CaCO_3 + SO_2 + 2H_2O + \frac{1}{2}O_2 \rightarrow CaSO_4 \bullet 2H_2O(s) + CO_2$	Ref 1
Other Processes	Ammonia (NH_3)	no	$SO_2 + 2NH_3 + H_2O + \frac{1}{2}O_2 \rightarrow (NH_4)2SO_4$	Ref 1,5
	Activated carbon	no	$SO_2 + H_2O + \frac{1}{2}O_2 \rightarrow H_2SO_4$	Ref 6

Ref 1. Srivastava, R., W. Jozewicz, and C. Singer, 2001.

Ref 2. Srivastava, R. and W. Jozewicz , 2001.

Ref 3. Davenport, 2006. Ref 4. Benson, 2003. Ref 5. He, 2002.

Ref 6. EPA, 2005. Ref 7. Shand, 2009. Ref 8. Maziuk, 2002.

3. Coal-Fired EGU CO_2 Control Technologies

The development of effective and commercially viable CO_2 control technologies for coal-fired EGUs is receiving widespread attention from the utilities, technology providers, and government agencies. Some CO_2 control technologies are still in the research and development phase and are not yet ready for commercial application. Other CO_2 control technologies are being demonstrated at larger scales and are progressing towards commercial viability. This remains an active area of research and new projects and technology advances are reported routinely. The discussions of CO_2 mitigation technologies and options presented in this section are based on the development status of a given technology as described in publicly available information as of May 2010.

3.1. Coal-Fired EGU CO_2 Emissions Control Approaches

A number of technologies lowering CO_2 emissions from coal-fired EGUs are currently commercially available or under development. These control measures use one of two basic approaches to reduce the amount of CO_2 released to the atmosphere: 1) by reducing the amount of fuel used (and the amount of CO_2 formed) by improving the energy efficiency of the electrical generation process, or 2) by separating the CO_2 for long-term storage using carbon capture technology.

3.1.1. Efficiency Improvements

When the efficiency of the power generation process is increased, less coal is burned to produce the same amount of electricity. This provides the benefits of lower fuel costs and reduced air pollutant emissions (including CO_2). A number of energy efficiency technologies are available for application to both existing and new coal-fired EGU projects that can provide incremental step improvements to the overall thermal efficiency. The energy efficiency technologies with the potential to achieve the greatest improvements in electric power generation efficiency involve EGU design, equipment selection, and cost decisions that are typically incorporated during the planning and engineering design phases for a new EGU project.

3.1.2. Carbon Capture and Storage

Carbon capture and storage (CCS) involves the separation and capture of CO_2 from flue gas, or syngas in the case of IGCC. It also requires pressurization of the captured CO_2, transportation via pipeline if necessary, and injection and long-term geologic storage. Several different technologies, at varying stages of development, may be considered for the CO_2 separation and capture. Some have been demonstrated at the slip-stream or pilot-scale, while many others are still at the bench-top or laboratory stage of development.

Development of commercially viable processes for capturing CO_2 from EGUs is being funded by U.S. DOE, electric utility companies, and other organizations. These processes typically use solvents, solid sorbents, and membrane-based technologies for separating and capturing CO_2. Amine-based solvent systems are in commercial use for scrubbing CO_2 from industrial flue gases and process gases. However, solvents have yet to be applied to removing the large volumes of CO_2 that would be required for a coal-fired EGU. Solid sorbents can be used to capture CO_2 through chemical adsorption, physical adsorption, or a combination of the two effects. Membrane-based capture uses permeable or semi-permeable materials that allow for the selective transport/separation of CO_2. Oxy-combustion uses high-purity oxygen (O_2) instead of air to combust coal, producing a highly concentrated CO_2 stream that does not require a separation/capture step.

Once the CO_2 is captured, it is transported, if necessary, and stored. Geologic formations such as oil and gas reservoirs, unmineable coal seams, and underground saline formations are potential options for long-term storage. Basalt formations and organic rich shales are also being investigated for potential use as storage. Beneficial reuse (e.g., enhanced oil recovery or carbonation) is a potential alternative to strict storage that provides potential revenue to offset a portion of the CCS costs.

One recent study prepared for the U.S. DOE by the Pacific Northwest National Laboratory (PNNL, 2009) evaluated the development status of various CCS technologies. The study addressed the availability of capture processes; transportation options (CO_2 pipelines); injection technologies; and measurement, verification, and monitoring technologies. The study concluded that, in general, CCS is technically viable today. However, full-scale carbon separation and capture systems have not yet been installed and fully integrated at an EGU. The study also did not address the cost or energy requirements of implementing CCS technology. For up-to-date information on Department of Energy's National Energy Technology Laboratory's (NETL) Carbon

Sequestration Program go to the NETL web site at: http://www.netl.doe.gov/technologies/carbon_seq/.

In 2010, an Interagency Task Force on Carbon Capture and Storage was established to develop a comprehensive and coordinated federal strategy to speed the commercial development and deployment of CCS technologies. The Task Force is specifically charged with proposing a plan to overcome the barriers to the widespread, cost-effective deployment of CCS within 10 years, with a goal of bringing 5 to 10 commercial demonstration projects online by 2016. As part of its work, the Task Force prepared a report that summarizes the state of CCS and identified technical and non-technical barriers to implementation. For additional information on the Task Force and its findings on CCS, go to: http://www.epa.gov/climatechange/policy task force.html. Because the development status of CCS technologies and their applicability to coal-fired EGUs are thoroughly discussed in the Task Force report, there will be no further discussion in this document.

3.2. Efficiency Improvements for Existing Coal-fired EGU Projects

Numerous efficiency improvements can be applied to coal-fired EGUs to increase thermal efficiency of power production (NETL, 2008, Sargent & Lundy, 2009; U.S. DOE, 2009; U.S. DOE, 2010). One specific example is the NETL study, which conducted a literature review of published articles and technical papers identifying potential efficiency improvement techniques applicable to existing coal-fired EGUs. Efficiency improvements can be expressed in different formats; they may be reported as an absolute change in overall efficiency (e.g., a change from 40% to 42% represents a 2% absolute increase). They may also be presented as the relative change in efficiency (e.g., a change from 40% to 42% is a relative change in efficiency and fuel use of 5%). The relative change in efficiency is the most consistent approach, since it corresponds to the same change in heat rate.

A summary of the findings from the NETL study is presented in **Exhibit 3-1**. The efficiency percentages were converted to a common basis so that all of the data could be compared. All of these improvements could not necessarily be implemented at each coal-fired EGU because of site-specific factors.

3.3. Efficiency Improvements for New Coal-Fired EGU Projects

3.3.1. Steam Cycle

The theoretical maximum achievable thermal efficiency achievable by an EGU using the Rankine cycle regardless of the technologies used is approximately 63% because of thermodynamic limitations and energy losses that cannot be recovered. Existing coal-fired EGUs using the Rankine cycle operate well below this limit. If the energy input to the cycle is kept constant, increasing the pressures and temperatures for the water-steam cycle will increase the output and the overall efficiency. However, a practical limitation to the higher pressure and temperatures that can be achieved in a boiler is the availability of boiler materials that can withstand these elevated conditions over an acceptable service life. The majority of existing PC-fired EGUs have subcritical boilers. Subcritical boilers typically operate at pressures of 2,400 psi (17 MPa) and at temperatures between 1,000 to 1,050°F (540 to 570°C). However, subcritical boilers can be designed to operate at steam pressures as high as 3,200 psi (22 MPa) and steam temperatures as high as 1,050°F (570°C).

The use of materials that can withstand the high-temperature and pressure of supercritical steam conditions allows for substantial improvements in efficiency for EGUs. "Supercritical" is a thermodynamic term describing the state of a substance where there is no clear distinction between the liquid and the gaseous phase (i.e., they are a homogenous fluid). Technically, the term "boiler" should not be used for a supercritical pressure steam generator, as no "boiling" actually occurs in this device, but it is common practice to use the term "Benson boiler." Supercritical EGUs typically use steam pressures of 3,500 psi (24 MPa) and steam temperatures of 1,075°F (580°C). However, supercritical boilers can be designed to operate at steam pressures as high as 3,600 psi (25 MPa) and steam temperatures as high as 1,100°F (590°C). Above this temperature and pressure the steam is sometimes called "ultra-supercritcal".

For a supercritical boiler, the feed water enters the boiler, is converted to steam, and is passed directly to the steam turbine (a supercritical boiler does not have a steam drum as shown in Exhibit 2-3 for a subcritical boiler PC-fired EGU configuration). Because the water-steam cycle medium is a single phase fluid with homogeneous properties, there is no need to separate steam from water in a drum. Supercritical boilers operate as once-through boilers in which the water and steam generated in the furnace waterwalls passes through only once. This eliminates the need for water/steam separation in drums during

operation, and allows a simpler separator to be employed during start-up conditions. Because these units do not have thick-walled steam drums, their start-up times are quicker, further enhancing efficiency and plant economics. Due to the availability of steam turbines that are designed for supercritical steam conditions, supercritical applications are presently limited to facilities of approximately 200 MWe gross output or more. Supercritical boilers are a well-established technology, and over 500 supercritical plants are currently operating worldwide (VGB, 2008).

The majority of EGUs have a single reheat cycle where the steam is first passed through the high pressure portion of the steam turbine and is then reheated in the boiler prior to passing through the remainder of the turbine. This process increases the efficiency of the EGU without increasing the maximum steam temperature. An additional steam cycle improvement that further increases efficiency is the use of a double reheat cycle, which reduces fuel use by approximately 1.5% compared to a similar EGU using a single reheat cycle (Retzlaff, 1996). The efficiency benefits of using a double reheat cycle have been recognized since the 1960s. However, the additional cost of a double reheat cycle has made a single reheat cycle typical for the majority of EGUs in the U.S.

To establish performance and cost baselines for analyzing EGU technology, NETL funded an independent assessment of the cost and performance of fossil energy power systems. The assessment specifically includes PC-fired boilers, IGCC, and natural gas-fired combined cycle systems in a consistent technical and economic manner reflecting market conditions for plants starting operation in 2010 (NETL, 2007). Performance and cost estimates were prepared for each configuration with and without CO_2 CCS. The Total Plant Cost (TPC) and Operation and Maintenance (O&M) costs for each of the cases in the study were estimated in January 2007 dollars and assumed plant construction on a generic site. The costs do not include owner cost and additional costs for special site-specific considerations at a given site.

For the PC-fired systems, the NETL study included a comparison of subcritical (2,400 psig/1,050°F/1,050°F) and supercritical (3,500 psig/1,100°F/1,100°F) PC-fired EGUs, each rated at nominal 550 MWe net capacity and firing Illinois No. 6 bituminous coal. A summary comparing the results for the subcritical unit versus the supercritical unit (without CO_2 CCS) is presented in Exhibit 3-2. The analysis shows an efficiency (HHV) of 36.8% for the subcritical boiler compared to 39.1% for the supercritical boiler.

Exhibit 3-1. Existing coal-fired EGU efficiency improvements reported for actual efficiency improvement projects

Efficiency Improvement Technology	Description	Reported Efficiency Increase[a]
Combustion Control Optimization	Combustion controls adjust coal and air flow to optimize steam production for the steam turbine/generator set. However, combustion control for a coal-fired EGU is complex and impacts a number of important operating parameters including combustion efficiency, steam temperature, furnace slagging and fouling, and NOX formation. The technologies include instruments that measure carbon levels in ash, coal flow rates, air flow rates, CO levels, oxygen levels, slag deposits, and burner metrics as well as advanced coal nozzles and plasma assisted coal combustion.	0.15 to 0.84%
Cooling System Heat Loss Recovery	Recover a portion of the heat loss from the warm cooling water exiting the steam condenser prior to its circulation thorough a cooling tower or discharge to a water body. The identified technologies include replacing the cooling tower fill (heat transfer surface) and tuning the cooling tower and condenser.	0.2 to 1%
Flue Gas Heat Recovery	Flue gas exit temperature from the air preheater can range from 250 to 350°F depending on the acid dew point temperature of the flue gas, which is dependent on the concentration of vapor phase sulfuric acid and moisture. For power plants equipped with wet FGD systems, the flue gas is further cooled to approximately 125°F as it is sprayed with the FGD reagent slurry. However, it may be possible to recover some of this lost energy in the flue gas to preheat boiler feedwater via use of a condensing heat exchanger.	0.3 to 1.5%
Low-rank Coal Drying	Subbituminous and lignite coals contain relatively large amounts of moisture (15 to 40%) compared to bituminous coal (less than 10%). A significant amount of the heat released during combustion of low-rank coals is used to evaporate this moisture, rather than generate steam for the turbine. As a result, boiler efficiency is typically lower for plants burning low-rank coal. The technologies include using waste heat from the flue gas and/or cooling water systems to dry low-rank coal prior to combustion.	0.1 to 1.7%
Sootblower Optimization	Sootblowers intermittently inject high velocity jets of steam or air to clean coal ash deposits from boiler tube surfaces in order to maintain adequate heat transfer. Proper control of the timing and intensity of individual sootblowers is important to maintain steam temperature and boiler efficiency. The identified technologies include intelligent or neural-network sootblowing (i.e., sootblowing in response to real-time conditions in the boiler) and detonation sootblowing.	0.1 to 0.65%

Efficiency Improvement Technology	Description	Reported Efficiency Increasea
Steam Turbine Design	There are recoverable energy losses that result from the mechanical design or physical condition of the steam turbine. For example, steam turbine manufacturers have improved the design of turbine blades and steam seals which can increase both efficiency and output (i.e., steam turbine dense pack technology).	0.84 to 2.6

Source: NETL, 2008

[a] Reported efficiency improvement metrics adjusted to common basis by conversion methodology assuming individual component efficiencies for a reference plant as follows: 87% boiler efficiency, 40% turbine efficiency, 98% generator efficiency, and 6% auxiliary load. Based on these assumptions, the reference power plant has an overall efficiency of 32% and a net heat rate of 10,600 Btu/kWh. As a result, if a particular efficiency improvement method was reported to achieve a 1% point increase in boiler efficiency, it would be converted to a 0.37 % point increase in overall efficiency. Likewise, a reported 100 Btu/kWh decrease in net heat rate would be converted to a 0.30% point increase in overall efficiency.

Exhibit 3-2. Summary of NETL performance, cost, and CO_2 emissions comparison analysis for nominal 550 MWe PC-fired EGU burning bituminous coal by steam cycle

Parameter	Bituminous Coal-Fired EGU	
	Subcritical Boiler	Supercritical Boiler
Gross Power Output (kWe)	583,315	580,260
Auxiliary Power Requirement (kWe)	32,780	30,110
Net Power Output (kWe)	550,445	550,150
Coal Flow Rate (lb/hr)	437,699	411,282
HHV Thermal Input (kW) Net Plant	1,496,479	1,406,161
HHV Net Efficiency (%)	36.8%	39.1%
Total Plant Cost ($ x 1,000)[a]	$852,612	$866,391
Total Plant Cost ($/kW)[a]	$1,549	1,575
Levelized Cost of Electricity (mills/kWh)[b]	64.0	63.3
CO2 Emissions (ton/hr)	519.1	487.6
CO2 Emissions (ton/year)	3,864,884	3,632,301
CO2 Emissions (lb/MMBtu)	203	203
CO2 Emissions (lb/MWh gross output)	1,780	1,681
CO2 Emissions (lb/MWh net output)	1,886	1,773

Source: NETL, 2007. Exhibit ES-2.

[a] The NETL costs are presented as "overnight costs" in January 2007 dollars and do not include escalation, owner's costs, taxes, site specific considerations, labor incentives, etc.

[b] 10 mills are equivalent to 1 U.S. cent.

Continuing research and advances in metallurgy have allowed the development of supercritical boilers capable of operating at increasingly higher temperatures and pressures, achieving increasingly higher efficiencies. Ultra-supercritical (USC) boilers designed to operate at steam conditions in excess of 4,500 psi can potentially operate at efficiencies approaching 50% (HHV). Steam temperature and pressure selection for boilers depends in part upon fuel corrosiveness, and research is focused on the development of new materials for boiler tubes and high alloy steels that minimize corrosion. There are potential concerns that temperatures above 1,100°F (590°C) while firing high-sulfur coal (such as Illinois No. 6) would result in an exponential increase of the material degradation of the highest temperature portions of the superheater and reheater due to coal ash corrosion. This could require pressure parts replacement outages every 10 to 15 years. The availability and reliability of materials required to support the elevated temperature environment for high sulfur or chlorine applications, although extensively demonstrated in the laboratory, has not been fully demonstrated commercially (NETL, 2007). Additional factors that could limit steam temperatures and pressures are the maximum values specified in the American Society of Mechanical Engineers (ASME) Boiler and Pressure Vessel Code, Section I: Power Boilers. A developer deviating from this code could have difficulty acquiring insurance, or be out of compliance with specific state code requirements.

The commercial use of ultra-supercritical technology has historically been, and continues to be, prevalent in countries outside the U.S., such as Denmark, Germany, and Japan. Ultra-supercritical boilers burning various coal ranks are being widely deployed throughout the world. Construction of the first modern ultra-supercritical EGU built in the U.S. began in 2008 at the Southwest Electric Power Company's John W. Turk, Jr. Power Station near Texarkana, AR. This 600 MWe PC-fired facility will burn PRB subbituminous coal and is scheduled to begin operation in late 2012. Examples of ultra-supercritical PC-fired EGUs often cited as representing the currently highest efficiency operating coal-fired EGUs in the world include:

- 384 MWe ultra-supercritical PC-fired EGU for the European Vattenfall company's Nordjylland Power Station Unit 3 located near Aalborg, Denmark (Vattenfall, 2006). This power station began operation in 1998 and is a combined heat and power (CHP) facility that generates electricity and produces heat for a district heating system. The unit burns imported bituminous coal and uses seawater for cooling. Reported overall unit efficiency is 47% LHV or 45.3% HHV (International Energy Agency [IEA] Clean Coal Centre, 2007).

- 965 MWe ultra-supercritical PC-fired EGU for the German RWE Power company's Niederaussem Power Station Unit K located near Cologne, Germany (RWE Power, 2004). This unit burns lignite with a 51 to 58% moisture content and started operation in 2002. Reported unit operating efficiency is 43.2% LHV or 37% HHV (IEA Clean Coal Centre, 2007).
- Two 600 MWe ultra-supercritical PC-fired EGUs for the J-POWER (Electric Power Development Co., Ltd.) Isogo Thermal Power Station New Unit 1 and New Unit 2 located in Yokohama, Japan (J-POWER, 2009). These units burn domestic and imported bituminous coal and use seawater for cooling. New Unit 1 started operation in 2002 and has a reported unit operating efficiency of 42% LHV or 40.6% HHV (IEA Clean Coal Centre, 2007). New Unit 2 started commercial operation in July 2009. The New No. 2 Plant, has a higher efficiency due to the boosting the reheat steam temperature 18°F higher than the New Unit No. 1 to 1,148°F (J-POWER, 2009).
- 450 MWe ultra-supercritical PC-fired EGU for the Capital Power Corporation's Genesee Power Station Unit 3 located near Edmonton, Alberta, Canada (Peltier, 2005). This unit burns Alberta subbituminous coal. This unit began operation in 2005 and has a reported unit operating net efficiency or 41% LHV or 39.6% HHV (IEA Clean Coal Centre, 2007).

As part of NETL's performance and cost baseline analysis for electricity production (NETL, 2007), IGCC EGUs were analyzed for the General Electric Energy (GEE), ConocoPhillips (CoP), and Shell coal gasification processes. The IGCC cases have different gross and net power outputs than the PC-fired EGU cases because of the combustion turbine size constraint. The advanced F-class turbine used to model the IGCC cases comes in a standard size of 232 MWe when operated on syngas. Each IGCC case uses two combustion turbines for a combined gross output of 464 MWe. Additional electrical output is generated by steam turbines, with steam from the HRSGs extracting heat from the combustion turbine exhaust. Although the two combustion turbines provide 464 MWe gross output in all cases, the overall combined cycle gross output ranges from 742 to 770 MWe. The net outputs range from 623 to 640 MWe depending on the gasification process. A summary comparison of the results for the three gasification processes without CO_2 CCS is presented in

Exhibit 3-3. Although the efficiency of the combined cycle block is approximately 50% efficient in converting the syngas to electricity, parasitic loads of the gasification process lower the net efficiency for the IGCC to 38.2 to 41.1% HHV.

Exhibit 3-3. Summary of NETL performance, cost, and CO_2 emissions comparison for an IGCC power plant by gasification process

Parameter	Gasification Process		
	GEE Radiant	CoP E-GasTM	Shell
Gross Power Output (kWe)	770,350	742,510	748,020
Auxiliary Power Requirement (kWe)	130,100	119,140	112,170
Net Power Output (kWe)	640,250	623,370	635,850
Coal Flow Rate (lb/hr)	489,634	463,889	452,620
HHV Thermal Input (kW) Net Plant	1,674,044	1,586,023	1,547,493
HHV Efficiency (%)	38.2%	39.3%	41.1%
Total Plant Cost ($ x 1,000)[a]	1,160,919	1,078,166	1,547,483
Total Plant Cost ($/kW)[a]	1,813	1,733	1,977
Levelized Cost of Electricity (mills/kWh)[b]	78.0	75.3	80.5
CO2 Emissions (ton/hr)	561.9	539.1	527.1
CO2 Emissions (ton/year)	3,937,728	3,777,815	3,693,990
CO2 Emissions (lb/MMBtu)	197	199	200
CO2 Emissions (lb/MWh gross output)	1,459	1,452	1,409
CO2 Emissions (lb/MWh net output)	1,755	1,730	1,658

Source: NETL, 2007. Exhibit ES-2.

[a] The NETL costs are presented as "overnight costs" in January 2007 dollars do not include escalation, owner's costs, taxes, site specific considerations, labor incentives, etc.

[b] 10 mills are equivalent to 1 U.S. cent.

3.3.2. Coal Drying

Low-rank coals (lignite and subbituminous) are often utilized because the low cost per unit of heat input relative to bituminous coal and the low sulfur content. However, a major disadvantage of low-rank coals is their high moisture content, typically 25 to 40%. When this coal is burned, considerable energy is required to vaporize and heat the moisture, thus raising the heat rate of the EGU and lowering its efficiency. As fuel moisture decreases, the heating value of the fuel increases so that less coal needs to be fired to produce the same amount of electric power. Drier coal is also easier to handle, convey, and pulverize – reducing the burden on the coal-handling system. In addition, an EGU boiler designed for dried coal is smaller and has lower capital costs than a comparable EGU designed to burn coal that has not been dried. The pre-combustion drying of low-rank coals can improve the overall efficiency and several advanced coal drying technologies are or nearly are commercial available.

Great River Energy developed a coal drying technology for low-rank coals in partnership with the U.S. DOE as part of the DOE's Clean Coal Power Initiative (U.S. DOE, 2007). The technology has been successfully demonstrated on a PC-fired boiler burning lignite at the utility's Coal Creek Station in Underwood, ND. The technology is now commercially offered under the trade name DryFiningTM (U.S. DOE, 2010b). The DryFiningTM process passes warm cooling water from the steam turbine exhaust condenser through an air heater where ambient air is heated before being sent to a fluidized bed coal dryer. The dried coal leaving the fluidized bed is sent to a pulverizer and then to the boiler. Air leaving the fluidized bed is filtered before being vented to the atmosphere. In addition to using power plant waste heat to reduce moisture, DryFiningTM also segregates particles by density. This means a significant amount of higher density compounds containing sulfur and mercury can be sorted out and returned to the mine rather than utilized in the boiler. The end result is that more energy can be extracted from the coal while simultaneously reducing emissions of mercury, sulfur dioxide, and NOX. At the Coal Creek Station, the process increased the energy content of the lignite from 6,200 to 7,100 Btu/lb, thereby resulting in a decrease in the fuel input into the boilers by 4% and a corresponding decrease in CO_2 emissions gains in overall efficiency of 2 to 4% are reported for the process.

RWE Power in Germany is also developing a fluidized bed drying technology for lignite, called WTA (RWE Power, 2009). A fundamental difference between the two drying processes is the WTA process first mills then dries the lignite while the DryFining™ process first dries then mills the lignite. A prototype commercial-scale drying plant using the WTA process began operation in 2009 at the utility's Nederaussem Power Station site. For the WTA process, lignite is first milled to a fine particle size by hammer mills in direct series with a two-stage fluidized-bed dryer. The dried fuel exiting the stationary bed is separated from the gas stream and mixed with coarser lignite solids collected from the bottom of the dryer bed and then fed directly to the boiler. The heat needed for the drying of the fuel is supplied by external steam, which is normally taken from the turbine with the heat transfer taking place in tube bundles located inside the bed. Based on the development work completed to date of the WTA technology, the net gain in cycle efficiency is reported to be on the order of 4 percentage points, depending on the moisture content of the raw coal and the final moisture of the dried lignite.

Several other coal drying technologies are in ongoing development. One coal drying process being developed by DBAGlobal Australia Pty, Ltd., with the trade name Drycol process uses the controlled application of microwave radiation to dry coal (Graham, 2007). Coal feed stock is first separated into fine grade coal and one or more larger grades. The fine coal is loaded onto a conveyor and conveyed continuously through a microwave-energized heating chamber for drying. The fine grade coal is dried sufficiently so that when it is recombined with the larger grade coals, the moisture content of the aggregate coal is within a target moisture content range. Other coal drying technologies for low-rank coals in various stages of development include: 1) attrition milling of coal followed by air drying to produce a low-moisture coal product, 2) compressing heated, coarse crushed coal to squeeze water out , and 3) heating wet coal under pressure to approximately 480 to 570°F (APP, 2008).

To date, it has not been economic to dry subbituminous coal at the mine prior to transport to an EGU. In addition, subbituminous coal that has been dried results in increased coal dust, can spontaneously combust, and will reabsorb moisture during transport and storage. However, the development of more efficient drying technologies such as dryers using flue gas recirculation and briquetting of the dried coal to avoid spontaneous combustion and moisture reabsorption can improve the economics of upgrading low-rank coals. Descriptions of upgrading low rank coals are available from the White Energy Company (http://www.white energyco.com) and Evergreen Energy (http://www.evgenergy.com/). From an overall GHG perspective, the increased EGU efficiency and decreased transportation GHG emissions would have to be compared to the energy required to dry and process the coal at the mine.

3.3.3. Boiler Feedwater Heating and Hot-Windbox

The high-pressure liquid water entering the steam generator is called feedwater. A feedwater heater is an EGU component used to pre-heat water delivered to the boiler section. Thermodynamic optimization of this cycle is important to overall EGU efficiency. In a conventional EGU, the energy used to heat the feedwater is steam extracted between the stages of the steam turbine (see Exhibit 2-3). Therefore, approximately a quarter of the steam that would be used to perform expansion work in the turbine (and generate power) is not utilized for that purpose. However, using other heat sources for the feedwater heater avoids the need to extract steam from the turbine allowing the steam to be used for electric power generation and increases the output of the steam cycle and potentially lowers GHG emissions. This alternate heat source can either be from an integrated solar thermal energy source or from a combustion turbine. Examples of solar thermal energy used to augment the steam cycle at combined cycle facilities include the Martin Next Generation Solar Center in Florida and the proposed Green Energy Partners/Stonewall, LLC facility in Virginia. The first coal-fired power plant to integrate solar thermal technology is the Cameo generating station in Colorado. In addition, EPRI (Electric Power Research Institute) is currently evaluating adding solar thermal energy to the Escalante and Mayo coal-fired power plants. An example of combustion turbine integration for feedwater heating is the Kettle Falls Generating Station (Schimmoller, 2003). For coal-fired boiler systems optimized to accommodate the combustion turbine exhaust, the incremental fuel efficiencies would be expected be comparable with combined cycle generation (Escosa, 2009; Stenzel). Another potential approach to integrate the use of a combustion turbine with a coal-fired steam cycle is using the turbine exhaust directly in the boiler in a hot-windbox. This involves injection of the combustion turbine exhaust directly into the boiler windbox or primary air ducts to provide an oxygen source as well as a heat source.

3.4. Combined Heat and Power Plant

Coal-fired EGUs dedicated to electric power generation and using the latest commercially available advanced technologies will generally operate at overall net efficiencies of approximately 40%. Significant amounts of energy released by coal combustion are lost during the steam condensation segment of the Rankine cycle due to heat transfer into the cooling water. In Europe, electricity is commonly generated by facilities that serve as both electricity generators and thermal energy producers for the local town or city district heating system. These combined heat and power (CHP) facilities are also known as cogeneration facilities. Operating an electric power station in a CHP mode allows recovery of some of the heat that would otherwise be rejected into cooling water, improving the overall efficiency of energy utilization. In applying CHP to an existing or new EGU, the temperature of the cooling water is normally not high enough to meet the requirements for most district heating or industrial process applications. In these cases, steam would be extracted at an elevated pressure and temperature from an intermediate stage of the steam turbine and then used for district or process heating.

This results in a decrease in the total electric power generation from the EGU. However, the overall fuel efficiency of CHP is higher than if electricity and steam were generated separately.

Because electricity can be transmitted over long distances, electric power plants can be located in remote areas as well as urban areas. However, thermal energy cannot be effectively transported over extended distances. This limits the practicality of incorporating a CHP mode into many electric power plant designs. The EGU needs to be located in close proximity to either a district energy system or an industrial facility with a significant and steady thermal demand. There are a number of examples; however, where industrial facilities have collocated with existing or new coal power plants in order to have access to reliable, low cost steam:

- DuPont's titanium dioxide plant in Johnsonville, Tennessee, is located next to TVA's Johnsonville power plant and buys high pressure process steam from the 1,200 MW facility. The power plant is comprised of 10 coal-fired boiler steam turbine units; the DuPont plant uses steam extracted from Units 1 through 4. Providing steam to the DuPont facility at the required process pressures reduces overall output of the power plant by 50 MW.
- Blue Flint Ethanol in Underwood, North Dakota, is a 50 million gallon per year dry mill ethanol producer located next to Great River Energy's 1,160 MW Coal Creek lignite-fired power plant. Starting operations in 2007, the ethanol facility purchases approximately 100,000 pounds per hour of medium pressure steam extracted from the power plant.
- Goodland Energy Center in Goodland, Kansas, is a small 22 MW coal-fired power plant that is supplying steam to a 20 million gallon per year ethanol plant and a 12 million gallon per year biodiesel plant, both co-located with the power plant. The power plant and the ethanol plant both started construction in 2006.

3.5. Oxygen Combustion

Oxygen combustion (oxy-combustion, oxy-firing or oxy-fuel) is an emerging technology applicable to either new or existing EGUs. The advantage offered by this technology is its potential for CO_2 emissions control because it produces a concentrated (nearly pure) CO_2 exhaust gas stream that requires minimal post-combustion clean-up prior to compression, transportation, and injection for long term storage. The basic concept of oxy-combustion is to use a mixture of oxygen (or oxygen-enriched air) and recycled flue gas (containing mostly CO_2) in place of ambient air for coal combustion. The resulting flue gas contains primarily CO_2 and water vapor with smaller amounts of oxygen, nitrogen, SO_2, and NOX. Consequently, the flue gas can be processed relatively easily to further purify the CO_2 (if necessary) for use in enhanced oil or gas recovery or for geological storage.

An oxy-combustion power plant consists of an air separation unit (ASU), an EGU with O_2-blown combustion, and a CO_2 treatment unit. The conventional ASU is a cryogenic process that has a significant energy requirement. However, alternative oxygen separation methods are being researched for possible commercial scale development. These alternative methods include ion transport membranes (ITM), ceramic autothermal recovery, oxygen transport membranes, and chemical looping (UARG, 2008). Oxygen is mixed with recirculated flue gas to create a mixture of O_2 and CO_2 (and some H_2O) which is used as the source of combustion oxidant instead of ambient air. The absence of air nitrogen produces a flue gas stream with a high concentration of CO_2.

Several research institutes are focusing on laboratory- and pilot-scale testing of oxy-fuel combustion (Levasseur, 2009). Pilot test programs currently are being conducted for European Enhanced Capture of CO_2 (ENCAP) program and the Advanced Development of the Coal-Fired Oxyfuel Process with CO_2 Separation (ADECOS) program. Additional research and development programs are being conducted, including:

- A 30 MW oxy-firing pilot plant at the Schwarze Pumpe station in Spremberg, Germany. This plant is the first complete oxy-combustion unit that includes the integrated system from the air separation unit to the gas purification and compression systems. The CO_2 will be compressed and liquefied for storage experiments to be conducted.
- A 32 MW oxy-firing demonstration project in France retrofitting an existing boiler to natural gas oxy-combustion. The captured CO_2 will be transported through an approximately 19 mile long pipeline and stored in a depleted gas field in Lacq, South of France.
- A comprehensive test program using the 15 MW tangentially-fired Boiler Simulation Facility and 15 MW Industrial Scale Test Facility operated by Alstom Power, Inc., in Windsor, CT. Testing is being conducted to assess a broad range of oxy-combustion design options.

Project partners include the U.S. DOE, the Illinois Clean Coal Institute, and 10 electric utility companies.

4. Coal-Fired EGU Technology Alternatives Analysis

There is no one best available coal-fired EGU technology universally applicable to all EGU projects. The coal-fired EGU technology alternatives most suitable for a given project must be evaluated on a site-specific basis. An evaluation for a new facility would include the use of carbon capture and storage and the most efficient technologies (e.g., ultra-supercritical steam conditions, IGCC, pressurized fluidized bed), double steam reheat, coal drying, FGD technology, and CHP.

4.1. Site-Specific Coal-Fired EGU Technology Alternatives Analysis Example

This section summarizes the results for analyses prepared in support of an air permit application for a new 830 MW supercritical PC-fired EGU to be built by the Consumers Energy Company at their Karn-Weadock Generating Station in Bay County, Michigan. The EGU is designed to burn PRB subbituminous coal, but can also mix bituminous coal based on supply or price variations. A supercritical PC boiler with steam turbine throttle pressure of 3,805 psia and superheat and single reheat temperatures of 1,100 °F was proposed by the Consumers Energy Company.

The air permit application was initially submitted to the State of Michigan Department of Natural Resources and Environment (DNRE) in 2007. As an amendment to the permit application for the Consumers Energy project, the DNRE requested that Consumers Energy Company prepare and submit a top-down BACT analysis for IGCC plant as an alternative to building the supercritical PC-fired EGU. The supercritical PC-fired EGU air emissions controls included selective catalytic reduction (SCR), baghouse, wet flue gas desulfurization, hydrated lime injection, and activated carbon injection. The IGCC plant air emissions controls included SCR, Selexol or Sulfinol, black water handling equipment, and sulfur-impregnated activated carbon.

To estimate the IGCC plant capital costs, the permit applicant developed a premium factor for constructing an IGCC plant compared to a similar capacity supercritical PC-fired EGU based on information collected by the permit applicant including reported costs of other IGCC projects. This capital cost premium factor was then applied to the estimated costs for the supercritical PC-fired EGU. Additional cost estimates were made for fuel, water consumption, waste disposal, operation and maintenance, and pollutant allowance purchases required for each facility type. The reference IGCC plant capital costs were estimated to be approximately 24% higher than the supercritical PC-fired EGU. The analysis estimated that the cost of electricity generation from an IGCC unit would be approximately 37% higher for the IGCC unit than for the supercritical PC-fired EGU. The projected cost of generation for the supercritical PC-fired EGU was \$60/MWh compared to \$95/MWh for the IGCC plant. The cost estimates prepared for the analysis are summarized in Exhibit 4-1.

Exhibit 4-1. Supercritical PC-fired EGU and IGCC plant cost comparison Summary prepared for Consumers Energy EGU project

Cost Parameter[a]	800 MWe net Supercritical PC-fired EGU	800 MWe net IGCC Plant	Difference (IGCC vs. PC)
Capital Costs (\$)	\$2,671,916,111	\$3,526,039,934	\$854,123,823
Annualized Costs (\$/yr)	\$273,871,401	\$361,419,093	\$87,547,692
Fuel Cost (\$/yr)	\$117,624,738	\$130,679,946	\$13,055,208
Cooling Water Consumption Cost (\$/yr)	\$5,166,656	\$3,114,268	(-\$2,052,388)
Waste Disposal Cost (\$/yr)	\$316,937	\$201,042	(-\$115,895)
Operating and Maintenance Costs (\$/yr)	\$59,307,369	\$84,336,386	\$25,029,017
Total Annual Cost (\$/yr)	\$456,287,101	\$579,750,735	\$123,463,634
Annualized Cost of SO_2 Allowances (\$/yr)	\$2,442,975	\$610,531	(-\$1,832,445)
Annualized Cost of NO_x Allowances (\$/yr)	\$4,132,417	\$2,611,787	(-\$1,520,630)

Source: Consumers Energy Company, 2008.

[a] Costs are presented in September 2007 dollars and include owner's and financing costs.

Exhibit 4-2. Coal-fired EGU technology alternatives cost comparison summary prepared for Consumers Energy EGU project

Coal-Fired EGU Technology	40-yr. BusBar Cost Excluding CO2 Cost[a]	40-yr. BusBar Cost Including CO2 Cost[a]	Availability	Efficiency-Heat Rate (Btu/kWh)	CO2 Emission Rate (tons/MWh)
Supercritical PC-fired 830 MW	$97 per MWh	$133 per MWh	86% to 92%	9,134	0.94
Subcritical PC-fired	$101 per MWh	$136 per MWh	84% to 89%	9,407	0.97
Subcritical CFB boiler	$108 per MWh	$145 per MWh	87%	9,798	1.01
Supercritical CFB boiler	$108 per MWh	$144 per MWh	87%	9,508	0.98
IGCC plant	$128 per MWh	$162 per MWh	70% to 81%	9,490	0.93
Ultra-supercritical PC-fired	$98 per MWh	$133 per MWh	91%	9,019	0.93
Supercritical PC-fired with Carbon Capture and Storage (CCS)	$135 per MWh	$139 per MWh	86% to 91%	10,836	0.11
Supercritical PC-fired 500 MW	$104 per MWh	$140 per MWh	86% to 92%	9,134	

Source: Consumers Energy Company, 2009.

[a] Busbar costs is the cost to generate the power leaving the plant (beyond the generator but prior to the voltage transformation point in the plant switchyard) and include all plant fixed costs (including all costs associated with the capital investment), fuel costs, operating and maintenance costs, emissions costs, interconnection costs, and transmission system upgrade costs. A busbar cost excluding CO_2 costs assumes a CO_2 tax or cap-and-trade program has not been implemented. A busbar cost including CO_2 costs assumes a CO_2 tax cost of $22/ton beginning in 2012 and rising to $53/ton by 2025.

At the request of the State air permitting authority for the project, the permit applicant prepared and submitted a second electric generation alternatives analysis for the Consumers Energy project in June 2009 (Consumers Energy Company, 2009). As part of this alternatives analysis, a comparison of the various coal-fired EGU technologies was presented with respect to air emissions from coal combustion including CO_2. The ultra-supercritical PC-fired EGU has the lowest projected heat rate of the analyzed technologies. Of the technologies without carbon capture and storage, it also has the lowest GHG emissions rate. In December 2009, the Michigan Department of Environmental Quality issued the construction permit based on the proposed supercritical PC boiler. The results are summarized in Exhibit 4-2.

4.2. EPA GHG Mitigation Database

The EPA Office of Research & Development (ORD) is collecting information regarding CO_2 mitigation measures applicable to coal-fired EGUs for compilation in a publicly-accessible GHG Mitigation Database. Version 1 of this database is expected to be released to the public in late summer 2010. The database is a tool that provides information of both commercially available technologies, as well as emerging technologies that are being demonstrated at larger scales for commercial viability.

REFERENCES

Internet Web Site addresses cited for individual references were available as of the date of publication of this document.

Asai, Akihisu, et al. 2004. *System Outline and Operational Status of Karita Power Station New Unit 1 (PFBC)*. JSME International Journal, Series B. Vol. 47, No. 2, 2004. Pp. 193-199. Available at: <http://www.jstage.jst.go.jp/article/jsmeb/47/2/193/_pdf>.

Asia-Pacific Partnership on Clean Development and Climate (APP). 2008. *Brown Coal Drying Technologies*. APP Joint Meeting of Cleaner Fossil Energy and Power Generation and Transmission Task Forces. April 1, 2008. Available at: <http://www.asiapacificpartnership.org/pdf/CFE/meeting_melbourne/SurveyofBrownCoalDr yingTechnologies-Godfrey.pdf>.

Benson, Lewis, et al. 2003. *Control of Sulfur Dioxide and Sulfur Trioxide Using By-Product of a Magnesium-Enhanced Lime FGD System.* Presented at ICAC Forum '03 Multi-Pollutant Emission Controls & Strategies, Multi-Pollutant Emission Controls & Strategies, Nashville, TN. October 14-15, 2003. Available at: <http://www.carmeusena.com/files/files/techpapersreports/carmeuse_icac_20forum_2003.pd f>.

Benson, Lewis. *New Magnesium-Enhanced Lime Flue Gas Desulfurization Process*. Technical Paper Report, Carmeuse North America. Pittsburgh, PA. Available at: <http://www.carmeusena.com/files/files/TechPapers Reports/fgd_new_magnesium.pdf >.

Consumers Energy Company. 2008. *Amendment to Application No. 341-07 Consumers Energy Company – Bay Count Top-Down BACT Analysis for IGCC.* April 2008. Consumers Energy Company, Jackson, MI. Available at: < http://www.deq.state.mi.us/aps/downloads/permits/CFPP/2007/341-07/TopDown%20BACT%20Analysis%20for%20IGCC.pdf >.

Consumers Energy Company. 2009. *Balanced Energy Initiative: Electric Generation Alternatives Analysis***.** June 2009. Consumers Energy Company, Jackson, MI. Available at: < http://www.deq.state.mi.us/aps/downloads/permits/PubNotice/341- 07/Alternatives Analysis.pdf >.

Davenport, W.G., et al. 2006. *Sulfuric Acid Manufacture*. Southern African Pyrometallurgy. South African Institute of Mining and Metallurgy, Johannesburg, 5-8 March 5-8, 2006.

Escosa, Jesús M. and Luis M. Romeo. 2009. *Optimizing CO_2 Avoided Cost by Means of Repowering*. Applied Energy, 86 (2009) 2351–2358.

Foster Wheeler North America Corp. 2009. *Utility CFB Goes Supercritical - Foster Wheeler's Lagisza 460 MWe Operating Experience and New 600 - 800 MWe Designs*. Prepared by James Utt, Arto Hotta, and Stephen Goidich, Foster Wheeler North America Corp., Clinton, NJ, for presentation at Coal-Gen 2009, Charlotte, NC, August 19-21, 2009. Available at: <http://www.fwc.com/publications/tech_papers/files/TP_CFB_09_12.pdf>.

Graham, James. 2007. *Microwaves for Coal Quality Improvement: The Drycol Project.* DBAGlobal Australia, Milton Queensland, Australia. Presented at the SACPS/International Pittsburgh Coal Conference 2007, Johannesburg, South Africa, September 10-14, 2007.

Available at: <http://www.drycol.com/downloads/Drycol%20Paper%20AC PS1%20060608. pdf>.

He, Boshu, et al. 2002. *Temperature Impact on SO_2 Removal Efficiency by Ammonia Gas Scrubbing*, Energy Conversion and Management 44:2175-

2188. http://www.sciencedirect.com/science?_ob=ArticleURL&_ udi= B6V02-3WRC6DR3&_ user=775537&_origUdi=B6V2P-47CY482-G&_ fmt=high&_coverDate=07%2F31%2 F1998&_rdoc=1&_orig= article&_ acct=C0000429 38&_version=1&_urlVersion=0&_userid= 775537&md5 =e2354e5058cba8f383722059ffecd1 cd

Hong, B.D. and E.R. Slatick. 1994. *Carbon Dioxide Emission Factors for Coal*. Originally published in U.S. Energy Information Administration, Quarterly Coal Report, January-April 1994, DOE/EIA-0121(94/Q1). Washington, DC, August 1994, pp. 1-8. Available at: <http://www.eia.doe.gov/cneaf/coal/quarterly/co$_2$_article/co2.html>.

IEA Clean Coal Centre. 2007. *G8 Case Studies by the IEA Clean Coal Centre*. Prepared by Colin Henderson, IEA Clean Coal Centre, London, United Kingdom for presentation at Third International Conference on Clean Coal Technologies for Our Future, Sotacarbo Coal Research Centre, Carbonia Sardinia, Italy. May 15 -17, 2007. Available at:<http://www.iea-coal.org/publishor/system/component_view.asp?LogDocId=81704&PhyDocId=6360>.

J-POWER (Electric Power Development Co., Ltd.). 2009. *Annual Report 2009*. Electric Power Development Co., Ltd., Tokyo, Japan. October 2009. Available at: <http://www.jpower.co.jp/english/ir/pdf/2009.pdf>.

Kaplan, P. Ozge, et al. 2008. *Is It Better To Burn or Bury Waste for Clean Electricity Generation?*. Environmental Science & Technology, Vol. 43, No. 6, 2009, pp. 1711-1717.

Korhonen, S., et al. 2001. Methane and Nitrous Oxide Emissions in the Finnish Energy Production. Fortum Tech-4615.

Levasseur, A. A., et al.. 2009. *Alstom's Oxy-Firing Technology Development and Demonstration - Near Term $_{CO2}$ Solutions*. Presented at the 34th International Technical Conference on Clean Coal & Fuel Systems, Clearwater, FL. May 31 - June 4, 2009. Available at: < http://www.netl.doe.gov/technologies/coalpower/ewr/co$_2$/pubs/5290%20Alstom%20oxycombustion%20paper%20Clearwater%20jun09.pdf>.

Maziuk, John and John Kumm. 2002. *Comparison of Dry Injection Acid-Gas Control Technologies*. Presented at the 95th Annual Conference and Exhibition of the Air and Waste Management Association, Baltimore, MD. June 23-27, 2002.

Michigan Department of Natural Resources and Environment. 2009. PTI Application No. 341-07. December 29, 2009. Available at: <http://www.deq.state.mi.us/aps/downloads/permits/CFPP/2007/341-07/341-07.htm>.

National Energy Technology Laboratory (NETL), 2007. *Cost and Performance Baseline for Fossil Energy Plants, Volume 1: Bituminous Coal and Natural Gas to Electricity, Revision 1.* DOE/NETL-2007/1281. U.S. Department of Energy, National Energy Technology Laboratory, Pittsburgh, PA. August 2007. Available at: <http://www.netl.doe.gov/energyanalyses/pubs/Bituminous%20Baseline_Final%20 Report.pdf>,

National Energy Technology Laboratory (NETL), 2008. *Reducing CO_2 Emissions by Improving the Efficiency of the Existing Coal-fired Power Plant Fleet*, DOE/NETL-2008/1329. U.S. Department of Energy, National Energy Technology Laboratory, Pittsburgh, PA. July 23, 2008. Available at: <http://www.netl.doe.gov/energy-analyses/pubs/CFPP%20Efficiency-FINAL.pdf>.

National Energy Technology Laboratory (NETL), 2010a. *Overview of DOE's Gasification Program.* Presentation by Jenny Tennant, Technology Manager, Gasification, U.S. U.S. Department of Energy, National Energy Technology Laboratory, Pittsburgh, PA. January 25, 2010. Available at: <http://www.netl.doe.gov/technologies/coalpower/gasification/pubs/pdf/DOE %20Gasificatio n%20Program%20Overview%202010%2001-25%20v1v. pdf>.

National Energy Technology Laboratory (NETL), 2010b. *CCPI/Clean Coal Demonstrations Nucla CFB Demonstration Project, Project Fact Sheet.* U.S. Department of Energy, National Energy Technology Laboratory, Pittsburgh, PA. Accessed June 21, 2010. Available at: <http://www.netl.doe.gov/technologies/coalpower/cctc/summaries/nucla/nuclademo. html>.

National Energy Technology Laboratory (NETL), 2010c. *CCPI/Clean Coal Demonstrations Tidd PFBC Demonstration Project, Project Fact Sheet.* U.S. Department of Energy, National Energy Technology Laboratory, Pittsburgh, PA. Accessed June 21, 2010. Available at: <http://www.google.com/imgres?imgurl=http://www.netl.doe.gov/technologies/coalpo wer/cc tc/summaries/tidd/images/tidd_plant_bw.jpg&imgrefurl= http://www.netl.doe.gov/technologies/coalpower/cctc/summaries/tidd/tidddemo.html&usg=__PMP6wAdGSJtmGwsCiNKKCwryUc=&h=>

Pacific Northwest National Laboratory (PNNL), 2009. *An Assessment of the Commercial Availability of Carbon Dioxide Capture and Storage Technologies as of June 2009*, PNNL18520. Pacific Northwest National Laboratory, Richland, WA. June 2009. http://www.pnl.gov/main/publications/external/technical_reports/PNNL-18520.pdf>.

Peltier, Robert. 2005. *Genesee Phase 3, Edmonton, Alberta, Canada*. Power. July/August 2005. Available at: <http://www.epcor.ca/SiteCollection Documents/Corporate/pdfs/publications%20and%20newsletters/platts power05.pdf>.

Peltier, Robert. 2010. *Plant Efficiency: Begin with the Right Definitions*. Power. February 1, 2010. Available at: <http://www.powermag.com/gas/ Plant-Efficiency-Begin-with-the-RightDefinitions_2432.html>.

Retzlaff, Klaus M and W. Anthony Ruegger, 1996. *Steam Turbines for Ultrasupercritical Power Plants*. GE Power Generation, GER-3945A. Available at: <http://www.gepower.com/prod serv/products/tech docs/en /downloads/ger3945a.pdf>.

RWE Power. 2004. *Niederaussem Power Plant: A Plant Full Of Energy*. RWE Power Aktiengesellschaf, Essen, Germany. October 2004. Available at: <http://www.debriv.de/tools/download.php?filedata=1218532199.pdf&fil ename=Kraftwerk %20Niederaussem%20(englisch).pdf&mimetype=application/pdf>.

RWE Power. 2009. *WTA Technology: A Modern Process for Treating and Drying Lignite*. RWE Power Aktiengesellschaf, Essen, Germany. February 2009. Available at: < http://www.rwe.com/web/cms/ mediablob/ en/88166/data/183490/36906/rwe/innovations/po wer-generation/coal-innovation-centre/fluidized-bed-drying/download-wta-en.pdf>.

Sargent & Lundy. 2009. *Coal-Fired Power Plant Heat Rate Reductions*. Report Number SL-009597. Sargent & Lundy, Chicago, Il. January 22, 2009. Available at: <http://www.epa.gov/airmarkt/resource/docs/coalfired. pdf>.

Schimmoller, Brian, 2003. *Avista Kettle Falls*. Power Engineering. October, 2003. Available at: <http://www.powergenworldwide.com/ index/ display/articledisplay/189185/articles/powerengineering/volume-107/issue-10/departments/managing-the-plant/avista-kettle-falls.html>

Shand, Mark A., 2009. *A New Look at SO_2 Removal*. April 28, 2009. Available at: <http://www.risiinfo.com/technologyarchives/chemicalsnews>.

Srivastava, Ravi K, Wojciech Jozewicz and Carl Singer, 2001. *SO2 Scrubbing Technologies: A Review*. Environmental Progress, 20(4):219-28.

Srivastava, Ravi K and Wojciech Jozewicz. 2001. *Flue Gas Desulfurization: The State of the Art*, Journal of the Air & Waste Management Association, 51:1676-88.

Stenzel, William C., et al. *Repowering Existing Fossil Steam Plants*. December 11, 1997. Available at: <http://soapp.epri.com/press/ default.htm>.

Utility Air Regulatory Group (UARG). 2008. *A Review of Carbon Capture and Sequestration (CCS) Technology*, Prepared by J. Edward Cichanowicz, Saratoga, CA, December 2008.

U.S. Department of Energy (DOE), 1996. *Electric Utility Engineer's FGD Manual, Vol. 1--FGD Process Design*. U.S. Department of Energy, Office of Fossil Energy, Morgantown, WV. March 1996. Available at: < http://www.netl.doe.gov/technologies/coalpower/ewr/pubs/fgdmanual_vol1.pdf>.

U.S. Department of Energy (DOE). 2007. *Clean Coal Technology Power Plant Optimization Demonstration Projects Topical Report Number 25*. September 2007. Available at: < http://www.netl.doe.gov/technologies/coalpower/cctc/topicalreports/pdfs/topical25.pdf>

U.S. Department of Energy (DOE), 2009. *Technical Workshop Report: Opportunities to Improve the Efficiency of Existing Coal-Fired Power Plants*. Workshop sponsored by the U.S. Department of Energy and National Energy Technology Laboratory. Rosemont, IL. July 15-16, 2009. Available at: <http://www.netl.doe.gov/energyanalyses/pubs/NETL%20Power%20Plant%20Efficiency%20Workshop%20Report%20Final.pdf>.

U.S. Department of Energy (DOE), 2010. *Technical Workshop Report: Improving the Efficiency of Coal-fired Power Plants in the United States*. Workshop sponsored by the U.S. Department of Energy and National Energy Technology Laboratory. Baltimore, MD. February 24-25, 2010. Available at: <http://www.netl.doe.gov/energyanalyses/refshelf/detail.asp?pubID=306>.

U.S. Department of Energy (DOE). 2010b. *Innovative Drying Technology Extracts More Energy from High Moisture Coal*. Fossil Energy Techline. March 11, 2010. Available at: <http://fossil.energy.gov/news/techlines/2010/10006-CCPI Technology Goes Commercial.html>.

U.S. Energy Information Administration (U.S. EIA). 2008. *EIA-923 (Schedule 2) - Monthly Utility and Nonutility Fuel Receipts and Fuel Quality Data (2008 Final Data)*. U.S. Energy Information Administration, Office of Coal, Nuclear, Electric and Alternate Fuels, U.S. Department of Energy, Washington, DC 20585. Accessed June 21, 2010. Available at: <http://www.eia.doe.gov/cneaf/electricity/page/eia423.html>.

U.S. Energy Information Administration (U.S. EIA). 2010. *Electric Power Annual 2008*. DOE/EIA-0348(2008). U.S. Energy Information Administration, Office of Coal, Nuclear, Electric and Alternate Fuels, U.S. Department of Energy, Washington, DC 20585. January, 2010.

Available at: <http://www.eia.doe.gov/cneaf/electricity/epa/epa_sum.html >.

U.S. Environmental Protection Agency (EPA). 2001. *Database of information collected in the Electric Utility Steam Generating Unit Mercury Emissions Information Collection Effort*. OMB Control No. 2060-0396. U.S. Environmental Protection Agency, Office of Air Quality Planning and Standards, Research Triangle Park, NC. April 2001.

U.S. Environmental Protection Agency (EPA). 2005. *Multipollutant Emission Control Technology Options for Coal-fired Power Plants*. EPA. EPA-600/R-05/034. U.S. Environmental Protection Agency, National Risk Management Research Laboratory. Research Triangle Park, NC. March 2005. Available at: < http://www.epa.gov/airmarkt/resource/docs/multipreport2005.pdf>.

U.S. Environmental Protection Agency (EPA). 2006. *Environmental Footprints and Costs of Coat-Based Integrated Gasification Combined Cycle and Pulverized Coal Technologies*. EPA. EPA-600/R-06/006. U.S. Environmental Protection Agency, National Risk Management Research Laboratory. Research Triangle Park, NC. July 2006. Available at: <http://www.epa.gov/oar/caaac/coaltech/2007_01_epaigcc.pdf>.

Vattenfall. 2006. *Nordjylland Power Station: The World's Most Efficient Coal-Fired CHP Plant*. Vattenfall, A/S Copenhagen, Denmark. Available at: <http://www.vattenfall.dk/da/file/nordjyllanduk080121_7841588.pdf>.

VGB Powertech and Evonik Energy Services (2008). Supercritical and ultra supercritical technology. In Power Plant Performance Reporting and Improvement under the Provision of the Indian Energy Conservation Act: Output 1.1: Best practice performance monitoring, analysis of performance procedures, software and analytical tools, measuring instrumentation, guidelines or best practice manuals and newest trends. (Annexure XIV). Indo German Energy Programme. Retrieved from <http://www.emtindia.net/PowerPlantComponent/Output1.1/Output1.1.pdf>.

Weijuan, Yang, et al. 2007. Nitrous Oxide Formation and Emission in Selective Non-Catalytic Reduction. Energy Power Engineering China, 228-232.

INDEX

D

E

O

P

Q

R

S

T

U

V

W